THE COST AND EFFECTIVENESS OF AUTOMOTIVE EXHAUST EMISSION CONTROL REGULATIONS

ORGANISATION FOR ECONOMIC CO-OPERATION AND DEVELOPMENT

PARIS 1979

The Organisation for Economic Co-operation and Development (OECD) was set up under a Convention signed in Paris on 14th December 1960, which provides that the OECD shall promote policies designed:
— to achieve the highest sustainable economic growth and employment and a rising standard of living in Member countries, while maintaining financial stability, and thus to contribute to the development of the world economy;
— to contribute to sound economic expansion in Member as well as non-member countries in the process of economic development;
— to contribute to the expansion of world trade on a multilateral, non-discriminatory basis in accordance with international obligations.

The Members of OECD are Australia, Austria, Belgium, Canada, Denmark, Finland, France, the Federal Republic of Germany, Greece, Iceland, Ireland, Italy, Japan, Luxembourg, the Netherlands, New Zealand, Norway, Portugal, Spain, Sweden, Switzerland, Turkey, the United Kingdom and the United States.

© OECD, 1979
Queries concerning permissions or translation rights should be addressed to:
Director of Information, OECD
2, rue André-Pascal, 75775 PARIS CEDEX 16, France.

TABLE OF CONTENTS

Preface.. 7
Conclusions.. 8
Executive Summary... 11

1. INTRODUCTION 16
   1.1 Study Purpose................................... 16
   1.2 Study Scope..................................... 17

2. WORLD AUTOMOTIVE EMISSION STANDARDS - DEVELOPMENT AND CURRENT STATUS................................. 18
   2.1 Overview.. 18
   2.2 United States Emission Standards 20
   2.3 Japanese Emission Standards..................... 24
   2.4 ECE Emission Standards.......................... 26
   2.5 Emission Standards in Other Countries........... 29

3. COST OF EMISSION CONTROL............................ 31
   3.1 Introduction.................................... 31
   3.2 Initial Cost of Control......................... 32
   3.3 Effect of Emission Control on Fuel Economy..... 41
   3.4 Maintenance Cost................................ 47

4. PERFORMANCE OF EMISSION CONTROL SYSTEMS IN USE..... 50
   4.1 Introduction.................................... 50
   4.2 Current Control Systems Degradation............. 51
   4.3 Advanced Control System Durability.............. 53
   4.4 Reasons for High Degradation Rates.............. 53
   4.5 Options for Improving Emission Control Performance In-Use................................ 63

5. CONCLUSIONS... 74

REFERENCES.. 77

Appendix
   Emission Test Cycles Around the World.............. 81

LIST OF EXHIBITS

2.1. Review of Initiation Dates for Automotive Emission Control Programmes........................ 20
2.2. Conceptualisation of Cost vs Control Relationship... 21
2.3. Comparison of Test Cycles........................ 22
2.4. U.S. New Automobile Exhaust Emission Control History 24
2.5. Japanese New Automobile Exhaust Emission Control History.. 25
2.6. ECE New Automobile Exhaust Emission Control History. 27
2.7. Correlation of ECE Test to CVS-CH (ECE Vehicles).... 28
2.8. Current Automotive Emission Standards in Other Major Countries.. 29
3.1. U.S. Estimates of Cumulative Initial Costs of Emission Controls................................ 34
3.2. Estimated Control System Cost at Various Standards - Cost Optimal Scenario............................ 35
3.3. Estimated Control System Cost at Various Standards - Fuel Optimal Scenario............................ 35
3.4. Effect of Standards and Technology Development on System Cost.................................... 36
3.5. BPICA (ECE) Cost Data............................ 38
3.6. BPICA Estimates of Costs of Attaining Various Control Levels by Vehicle Weight.......................... 39
3.7. Comparison of CCMC and EPA Estimated Cost of Control Hardware.. 40
3.8. Impact of Various Emission Control Techniques on Fuel Economy.................................... 42
3.9. Per Cent Change in Fuel Economy from 1957-67 Average vs Inertia Weight................................ 43
3.10. Effect of Standards and Technology Development on Fuel Economy.................................... 43
3.11. Engine-out HC Emissions for Good Fuel Economy with no HC Emission Control............................ 44
3.12. Percentage Increase in Fuel Consumption in ECE Automobiles As a Result of Reduced Emission Standards.. 46

3.13. Effect of Japanese Emission Control Standards on Fuel Economy....................................... 47

3.14. Estimates of Lifetime Control Maintenance Costs..... 48

4.1. EPA Estimated Average In-use Emission Rates for Current Control Systems............................. 52

4.2. Estimates of the Field Performance of Emission Controls.. 52

4.3. EPA Estimated Average In-Use Emission Rates for Advanced Control Systems............................ 54

4.4. Incidence of Component Maladjust or Failure in 1975-76 California Vehicles........................ 55

4.5. Number of Vehicles that fail the Federal Standards by Standard for the Cities of Chicago, St. Louis, Washington and Phoenix.............................. 56

4.6. Frequency of Idle CO Maladjustment (Per cent) for 1975 Vehicles...................................... 57

4.7. Frequency of Failure for Selected Components Source Champion Spark Plug Company........................ 57

4.8. Degradation of Idle Carbon Monoxide with Vehicle Age. 58-59

4.9. Degradation of Idle Hydrocarbon with Vehicle Use.... 60-61

4.10. Relationship of Idle Screw Position to Idle % Carbon Monoxide and Idle RPM.............................. 63

4.11. Comparison of Owner Maintenance Responses with As-Received Emission Tests............................. 64

4.12. Comparison of Owner Maintenance Responses with As-Received Emissions................................. 64

4.13. Emission Control Maintenance Frequency vs Vehicle Model Year... 66

4.14 Inspection Test Options............................ 70

4.15. Comparative Annual Operating Costs................. 70

4.16. Initial Effectiveness of Emission Maintenance....... 72

4.17. Average Cost of Repair............................. 72

PREFACE

Controlling emissions from motor vehicles is an important element of air pollution control policies, especially in urban areas. This objective is usually achieved by the setting of emission standards. However, motor vehicle emission control devices often deteriorate with use. Various regulatory and economic instruments are in use or have been proposed to ensure maintenance of emission standards. This report examines those instruments, examines cost data related to that and gives emission control surveys.

The report - which has been prepared under the responsibility of the Secretary-General of OECD - arrives at a number of well defined conclusions to provide guidelines for Member countries on the most cost-effective strategies for air pollution control of motor vehicles over their lifetime. The study has been carried out in close co-operation with other international organisations.

CONCLUSIONS

Recognising that the motor vehicle is a major contributor to air pollution especially in urban and suburban areas.

Recognising that the air quality in many areas and cities of OECD is improving partly as a result of motor vehicle emission control systems and emission inspection programmes.

The following conclusions from the review of the current practices in motor vehicle emission control are presented.

1. In all areas of OECD, the ambient air quality improvements are limited by major decrease in the effectiveness in automobile emission control systems which occurs because of degradation of the system in consumer use.
2. The major causes of degradation are component durability of the control system in consumer use, improper maintenance, maladjustment by the service industry and outright tampering.
3. As a result, the current ECE legislation is not sufficient by the time the vehicle has accumulated 80,000 km and the United States 1975 standards are equivalent to the original 1973 standards.
4. Governments should initiate major action-oriented efforts nationally and internationally through established bodies who are currently active in motor vehicle policies to introduce restrictions on the adjustability, durability and/or maintenance requirements of critical emission components. It appears that these actions would be the most cost effective of all available options to reach the same environmental goal.
5. The cost analysis indicates that if component adjustability limits were imposed there is the possibility of no net cost because the original cost is expected to be offset by service cost savings by the owner.
6. Inspection/Maintenance programmes have had various degrees of success in OECD countries. In general, inspection/maintenance is a less effective strategy than durability requirements, adjustment limitations and/or tamper-proof requirements.

7. In certain critical air quality areas the imposition of inspection maintenance might be worthwhile even though it has a relatively low effectiveness-cost ratio as currently applied.
8. Both the Japanese and Americans have committed themselves to low levels of emissions by the 1980s. In taking this step, they will be adding $35-110 into the cost of the vehicles or approximately doubling the total control cost of the 1977 systems. The effectiveness of these controls could be in serious jeopardy if adjustability limits or controls are not included on the control hardware as these advanced systems are even more sensitive to engine parameters that are adjustable on present day automobiles than the current generation of controls.
9. Governments of OECD Europe which follow the ECE automobile emission control standards should also consider adding a durability test to the new vehicle compliance test such as that which is required in the United States or Japan.
10. Governments of OECD Europe should also be considering the application of evaporative emission control standards as this standard has been proven to be very cost-effective in both the United States and Japan.
11. Investigations have shown that the automotive service industry is in general not adequately trained nor adequately equipped with service manuals, diagnostic equipment and special tools for emission control system repair. Governments should consider formulating guidelines at an international level to require improvements and standardizations of information, procedures and equipment.
12. Periodic compulsory vehicle inspection and maintenance could also pressure the service industry to improve the quality of its work even on uninspected vehicles, and the manufacturers to improve the service information and equipment.
13. In both Japan and the United States, the phased introduction of emission controls (i.e. California before the entire United States, and in Japan on new models before existing models and imports) has appeared to have significant benefits both from the control reliability and cost points of view.
14. Similar benefits might be present if OECD Europe used one small market portion as their "test" of the control system before they are applied uniformly across all of Europe. This trail testing will develop unique European data which can be used by the manufacturers to optimise their final large production run vehicles.

15. Finally, the selection of control strategies, hardware and abatement timetable is very much a political decision which must by definition vary from country to country. Although the politics of the controls may vary, every attempt must be made to supply the technical analysts with data representative of the state-of-the-art so that every possibility of an uninformed and imprudent decision is excluded.
16. In view of this, in both the United States and Japan, major efforts to further quantify the durability of the catalyst control systems should be carried out. It should be stressed that the aim of these field programmes should not be to assess whether the cars meet the standards but to develop and implement systems to improve the effectiveness of the controls during the lifetime in use.
17. Similarly, to improve the European data base on motor vehicle emissions and control technology, it is suggested that government sponsored co-operative research and assessment programmes be initiated within the ECE (or another appropriate established international organisation).
18. The countries which are not part of the three major standards all are facing the difficult question of what their next standard step should be. It is suggested that because all the countries have a common problem that they pool their knowledge and resources and co-operatively search for the solution. With the initiative of these countries, it may be possible to accelerate the international harmonization of new vehicle test procedures, of durability, adjustability and tamper-proof requirements and of improved servicing information and equipment.

EXECUTIVE SUMMARY

Most developed nations have imposed automotive emission standards in order to reduce the atmospheric contamination in their major cities. A review of the history and current status of these regulatory programmes indicated that there are three sets of automotive emission standards to consider, namely those of the United States, Japan and the Economic Commission of Europe (ECE). A variety of other countries have current regulations which are not in any one of these blocs but instead are lagging behind the United States regulations by approximately three years. Each one of the three major standards are based on its own test method which makes direct comparison of standards difficult. Two studies on the comparability of standards have been identified. These studies show that the United States and Japan regulations are currently equal while the ECE regulations are approximately half to a third as stringent as the United States.

While there are technical problems with all the test methods, the most severe problems have been revealed in the ECE test. The specific problem areas are:

- the method of HC measurement is not exact enough for use at lower standard levels;
- the ECE test requires no vehicle durability driving and thus the control system may not be as durable as in the United States or Japan;
- the driving cycle has been shown to be non-representative of European driving patterns;
- the sampling procedure is not applicable for diesel exhaust measurement.

As the standards become more stringent, the "menu" of technology which can be employed changes. The selection of the exact combination of technologies (e.g. catalyst, EGR, spark control, etc.) is an economic decision which is up to the manufacturer. However, the type of control used is partially controlled by the standard setting process. For example, the current ECE standards may be approaching the dividing line between low (e.g. spark control, better engine tolerances, EGR) and high (e.g. catalyst, stratified charge) control technologies. With further reduction major hardware (or fuel) changes may be required which, although already applied in North America and Japan, will cause some dislocation in the economics of the industry.

In addition, if the control stringency is lowered down to or below the level of research technology (as opposed to production technology) some cost and fuel inefficiencies will be imposed. To circumvent this problem of application of new technology, both the United States and Japan allow for the gradual phasing in of advanced controls (in the United States by California's more stringent standards and in Japan by enforcing the standards first on new models only) thus allowing the manufacturers to prove the systems on a limited production before mass introduction. This procedure appears beneficial both from the manufacturers and customers cost point of view.

The majority of data on control costs were assembled from both European and United States sources. Both in the United States and in Europe there are substantial discrepancies between industry and government cost estimates. For the standards in Europe and the United States, the estimated initial costs per vehicle are:

Standard	Initial Cost	
	Government	Industry
ECE (77)	> $50	–
ECE (79)	≈ $50	$77
U.S.(77)	$33–$129	$140–$318
U.S.(80)	$68–$239	$238–$275

These discrepancies are caused by differing views on the components included in the cost accounting and their manufacturing and assembly costs as well as the vehicle size.

In addition to the initial production costs, there are fuel consumption and maintenance costs associated with levels of control. The fuel economy sensitivity to control level is very dependent upon the type and design of control hardware used. In many cases, the manufacturer tends to optimise for manufacturing costs and thus increases the fuel consumption, conversely a fuel optimised design generally increase costs. The greatest amount of analytical work on the relationship of control standards to fuel economy has been done in the United States where it has been concluded that current model year cars are superior in fuel economy to their precontrolled counterparts (normalizing for inertia weight). In addition, it appears that the manufacturers have the ability to optimise for fuel economy after several years at one standard. Thus, any new standard may impose only a transitory fuel penalty which is typically eliminated within five years.

At stringent control levels, the technology required to maintain fuel economy generally requires the use of a lead-free or low-lead gasoline in order that engine lead deposits are reduced and/or catalysts are not poisoned. The added cost of this premium fuel has been shown to be offset by maintenance cost reductions over the life of the car. Because of this switch in fuel, most maintenance cost estimates indicate lower costs with the more advanced systems. In addition, the requirement for more exact and durable temperature control, ignition system, fuel mixing and distribution systems and combustion chamber components all tend to reduce the servicing intervals.

With these product improvements, the lifetime maintenance costs are approximately the same or slightly higher than the initial hardware costs for the advanced control systems. In contrast, the maintenance costs are anywhere from three to four times the initial cost for the early United States and current European control systems. Therefore, the lifetime costs (initial plus maintenance costs) should be more thoroughly investigated by regulatory agencies to assess the true cost of control systems as it appears that added hardware cost is compensated by reductions in lifetime maintenance costs.

The costs of control, of course, must be compared to their effectiveness in reducing contaminant emissions. Here, it is deceiving to assess reductions by only referring to the regulated standards. After reviewing numerous emission control surveys in the United States, Canada, Japan, Sweden, and the ECE, the general conclusion that there is serious and rapid degradation of the emission rates was made. Using a five year old car as the indicator of the average lifetime emission rate, the cars currently in operation are operating with emission rates over twice the original standards. Thus, the 1975 United States cars actually achieved the average emission rate which was mandated in 1973. Similarly, the data from Europe indicate that the ECE standards have not had sufficient effect on reducing emission rates below the uncontrolled levels.

The most significant problems of high degradation occur with CO and HC. CO in particular is very susceptible to engine adjustment while HC emissions increases are caused by engine wear and poor ignition. From idle CO surveys results from many countries, the fact that per cent CO is well above manufacturer specifications at low vehicle use is apparent. This characteristic is independent of control technology as the fuel mixture and introduction systems are still adjustable even with the most advanced systems today. In the current and next generation of control systems in the United States and Japan, the additional problem of catalyst durability and poisoning is present. These systems are even more sensitive to engine adjustment than their predecessors and will at the same time be very sensitive to fuel composition.

While a portion of the current degradation problem is the fault of the original design durability, misfueling of vehicles and overt tampering of the control systems, the majority of cars are failing because of poor and/or inexact maintenance. The service industry at present has limited incentive to adjust and maintain cars for emissions and as a consequence opts to satisfy the customer on cost, driveability, and reliability. As the idle mixture can easily be used to "cure" minor problems with the car, it shows the highest incidence of maladjustment. Timing is the next most frequently maladjusted parameter. There are two solutions to the problem:

- more pressure can be placed on the servicing industry and the consumer to keep the car within specifications by imposing periodic vehicle inspections;
- restriction on the vehicle design to limit the adjustability of critical systems such as idle mixture and timing.

It appears from the limited cost data available on the latter strategy that it is the most cost-effective alternative. The use of inspection and maintenance is warranted only in those areas where air quality or vehicle demographics demand that more immediate and costly action be taken on all cars and not just the new additions to the fleet.

Due to the considerable decrease in emission control cost-effectiveness caused by control system degradation in consumer use, it is a conclusion of this report that major efforts be directed at reducing this problem by restricting the adjustability and increasing the durability of the control hardware. Although applicable in all jurisdictions, it is critical in the United States and Japan, that field performance of the control systems be given more priority. The development and imposition of adjustability, durability or allowable maintenance standards should be investigated before any additional lowering of new car standards is considered.

Additionally, it is concluded that a major review of the ECE regulations be initiated which would include reassessment of the certification procedures, the test method and gas analysis equipment. As it is apparent that Europe can still increase the stringency of its standards and maintain a high effectiveness-cost ratio while incurring no fuel penalty. This reduction could possibly be regionally phased-in (as in the United States and Japan) to cushion the industrial and consumer adjustment process.

Finally, more non United States data on cost and effectiveness of emission controls should be collected and published. Through this cross-fertilisation of data and ideas among nations whose basic environmental goals are similar, more enlightened and effective legislation, policies and programmes will be instituted to the benefit of governments, industry and the consumer.

The countries of the EEC are of the opinion that, with the efforts they have undertaken to date for motor vehicle emission control (including the ECE proposals for 1979), they have exhausted the possibilities for further emission reduction with existing engine concepts. Further substantial reductions will require new concepts such as, for example, stratified charge engines or catalytic convertors. The introduction of such concepts must take into account fuel consumption.

1. INTRODUCTION

1.1. STUDY PURPOSE

For over a decade, cars in North America have been subject to emission control standards. These standards, which first originated in California in the early 60s, have required major re-design of engines and have caused cost and energy repercussions in the automotive industry. Although the concern for automotive air pollution was first highlighted and abatement strategies first developed in the United States, concern for auto emission control has spread to many other countries of the world. Today, most developed countries have regulations on the allowed emission rates of Carbon Monoxide (CO), Hydrocarbons (HC) and Nitrogen Oxides (NOx). The control of cars on a world-wide level can be segmented into three major blocks, the United States, Japan and European Economic Community. While there are other countries which do not fall exactly into these three blocks, the major control activity has been confined to these three groups of countries and as such they represent the major portion of the analysis in this report. Each one of these emission control groups has developed abatement programmes to suit its own economic and political requirements. These differences have resulted in three distinct test cycles on which the standards are based and three distinct abatement timetables.

The differences in both schedule and test cycles cause some difficulty in comparing the various emission standards around the world. This study has attempted a rationalisation of the various test cycles in order to ascertain the relative stringency of the standards in each one of the countries. Little experimental work has been done on this subject of standard comparability and many of the results obtained so far are inconclusive. Further work on a suitable method of comparison of the various standards is needed. Despite these difficulties, it was possible in this study to group the various standards into relative levels of severity on the basis of the data at hand. The accuracy of this grouping is considered sufficient for the purposes of this study, however, the accuracy may not be sufficient for more detailed analysis of the various emission standards.

In addition to a discussion of the various standards around the world, the prime objective of this report was to ascertain the expected cost of control for various levels of stringency of emission control and the effectiveness of the control. It was not the goal of this report to estimate the potential air quality costs and benefits which would accrue from such control measures. The costs of control were developed from the literature and assume the imposition of several types of abatement strategies. These strategies include:

- imposition of new car emission standards only;
- in-use periodic vehicle inspection and maintenance;
- new car emission standards which entail a "low maintenance" requirements;
- new car emission standards with low adjustability regulations.

The measure of effectiveness of the strategies was the emission rates of vehicles in consumer use. Field surveillance data from various countries were used for the effectiveness measure.

Finally, the project was to suggest ways of optimising the control of emissions from automobiles based on the above analytical work. This was undertaken with the full realisation that country to country variations will always be present and in fact are useful in order that numerous policy options are tested for validity.

1.2. STUDY SCOPE

This study was designed to investigate only the cost implications of various exhaust emission control methods for automobiles. As such, it did not analyse either blowby or evaporative control systems nor did it evaluate light or heavy duty truck controls.

All costing was estimated on a car lifetime basis. The assumptions for the average life of the vehicle were:

- 160,000 km. total travel;
- 10 years of use.

As the cost of control varies with vehicle size, a range of costs was developed to indicate the impact on both the small European/Japanese size of vehicle and the larger United States size of automobile.

In the energy impact analysis, the effect of weight reduction was eliminated whenever the data allowed. All fuel improvements were calculated based on changes in fuel consumption and not fuel economy.

2. WORLD AUTOMOTIVE EMISSION STANDARDS DEVELOPMENT AND CURRENT STATUS

2.1. OVERVIEW

Following the linking of automotive exhaust to photochemical smog formation by Haagen-Smit almost 40 years ago, the car has come under increasing emission control in various parts of the world. The movement began in California in the late 1950s where the first enactment of regulations covering blowby exhaust from motor vehicles were established. During that time, there was a general realisation that the air quality in the major cities in the United States was deteriorating similarly to that seen in California. This prompted the United States Congress to establish the Clean Air Act in 1963. Through a series of amendments to this Act, more and more restrictions were placed on the car culminating in the 1970 emission standard amendment.(1) These amendments set a timetable for compliance with the legislated emission goal of a 90 per cent reduction from a 1970 automobile emission rate.

Following the United States regulation, this 90 per cent reduction goal was copied by numerous other governments. This trend can be noted in Exhibit 2.1. which shows the chronology of automotive regulation enactment. Although the United States has been the leader in the imposition of regulations, it is followed closely by Japan and then the European Economic Community. It can also be noted that, to the present time, only developed countries have instituted effective regulation of motor vehicle exhaust emissions. It can be expected that other less developed countries will adopt such emission control systems once the administrative and control costs are at a level which will not have any significant effect on their economic development and their technical abilities are increased.

As noted in Chapter 1, not all countries have adopted the same standards or test procedures. Thus, the individual standards that have been set cannot be directly compared until they have been converted into a common test procedure. This comparison is discussed in the next Section, 2.2. Additionally, the fact that there are three independent pollutants to be controlled and that various

Exhibit 2.1.

REVIEW OF INITIATION DATES FOR AUTOMOTIVE EMISSION CONTROL PROGRAMS

Country	Year	Country	Year
United States	1963	Sweden	1971
Japan	1966	United Kingdom	1972
ECE	1971	Australia	1972
Canada	1971	Finland	1975

countries have established control programmes which reduce these individual pollutants at different rates makes the comparison of the degree of control difficult. Simplistically, one could just sum the individual per cent reductions from the uncontrolled state for each one of the pollutants to obtain a percentage control value. However, this process would not account for the differential effect each pollutant has on cost and fuel economy. If fuel economy is considered to be a normal cost, then we are left with a four-dimensional array which can be conceptualised in a three-dimensional volume relationship such as shown in Exhibit 2.2. Even this level of complexity is too simplistic as there are other influences on the cost-effectiveness of controls which must be taken into account:

- type of technology used for control;
- lead-time for regulated emission reduction;
- size of vehicle;
- type and quality of available fuel;
- competency of repair industry.

With all of these parameters to be traded off, it is predictable that there is a multiplicity of opinion on the most effective control strategy. This has resulted in the range of standards currently applied to automobiles. It can be expected that this diversity will slowly diminish as technology development and energy restrictions select the optimal control techniques.

The Constant Volume Sampling-Cold/Hot test cycle (CVS-CH) is used as standard test cycle in this report. The selection of this cycle as opposed to the ECE or the Japanese cycle, was due to the vast amount of cost and fuel economy effect information that has been developed on this cycle. A description of this and other world test cycles is included in the Appendix of this report. The

CVS-CH cycle attempts to estimate the emission level of vehicles by having the vehicle complete a driving cycle on a dynamometer test stand which is representative of actual driving in the Los Angeles area. Many criticisms have been placed on the selection of this test cycle, however, in comparative studies of driving cycles in various other cities, there have been no truly significant exceptions to the percentage of acceleration, cruise, deceleration, idle and average speed in the United States. In general, the cycle represents a reasonable duty cycle which can be expected to be seen by most vehicles travelling in urbanised areas.

The test methods estimates the mass emission rate of the three pollutants using a technique known as the Constant Volume Sampling system. The test cycle and gas analysis train are sufficiently accurate and reproducible for most of the standards set in various world jurisdictions. There will, however, be reproductibility problems with this and other cycles when the emission control standards are set at very low levels. Because of these reproducibility problems, it may be necessary in the future to reassess the test method and come up with a test cycle or test conditions which reduce test to test variability. A comparison of the United States cycle to the other two major world test cycles (ECE and Japanese) is shown in Exhibit 2.3. which is extracted from the Appendix.

2.2. UNITED STATES EMISSION STANDARDS

The activity of California in the late 1950s stimulated United States congressional action which produced the Clean Air Act in 1963. This Act recognised the problem of air pollution and established research and development activities to attempt to control it. It was quickly evident that the Act was somewhat limited in its ability to control motor vehicle pollution and thus, in 1965, the Motor Vehicle Air Pollution Control Act was introduced. This Act set up a bureaucracy for setting the emission control standards and also included provision for anti-tampering legislation. Up to this point, no specific emission control standards had been set for the whole of the United States, although California had been active in establishing, in 1966, emission control standards on new motor vehicles.

In 1967, the Congress enacted the National Emission Standards Act which set emission standards for new cars produced after 1968. This Act also pre-empted any other state regulation on motor vehicles, excepting California. The problems that were uncovered by the institution of this Act and the perception of Congress that the industry was not doing its best effort to develop lower polluting vehicles led to the review of the 1965 and 1967 Acts.

Exhibit 2.2
CONCEPTUALIZATION OF COST vs CONTROL RELATIONSHIP

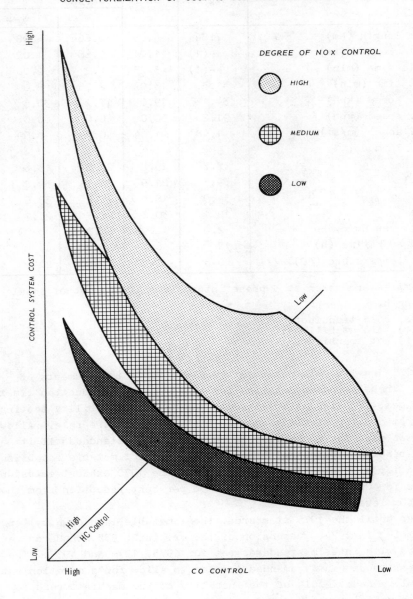

Exhibit 2.3.
COMPARISON OF TEST CYCLES

Parameter	Test			
	CVS-CH	ECE	JAPAN	
			10 Mode	11 Mode
Test length (km)	17.84	4.05	0.664	4.084
Cycle length (km)	-(1)	1.01	0.66	1.02
Cycle time (min)	-(1)	3.25	2.2	2.00
Test time (min)	31.3	13.0	2.25	8.00
Avg. speed (kph)	34.1	19.0	17.7	30.6
Max. speed (kph)	91.2	50.0	40.0	60.0
Max. accel (m/s^2)	1.48	1.04	0.79	0.69
Time in mode, %				
Cruise	7.7	29.2	23.6	13.3
Accel	39.3	21.5	24.5	34.2
Decel	34.9	18.5	25.2	30.8
Idle	18.1	30.8	26.7	21.7
Stops per km/cycle	2.0	2.97	0.30	0.09
Cold soak time (h)	12	6	-	6
Test temperature (°C)	20-30	20-30	20-30	20-30

1) Samples are taken at 3 phase intervals that are not of equal length.
   i) cold transient;
  ii) cold stabilized;
 iii) hot transient.

This review culminated in the Clean Air Act Amendments of 1970. These amendments called for a set emission reduction timetable to be met. It also established the concept of durability testing requirements for certification of the vehicle before sale. Allowance was made for a one-year suspension of the standards if it would benefit public health and welfare. The timetable established called for a 90 per cent reduction from the 1970 exhaust emission levels for CO and HC by 1975 and a 90 per cent reduction from the 1971 NOx levels by 1976.

Up until the 1975 standards, the timetable laid down by Congress were met. In 1974, suspension of the original 1975 standards was allowed with interim standards set for 1975, 1976 and 1977. The Congress subsequently amended the Act to allow for a more lenient schedule. As a result of the inability of the manufacturers to produce vehicles which could meet the original 1976 standards in 1973, Congress again reviewed the amendments and created additional amendments as of August, 1977.

Although the new amendments have not been reviewed for all their implications as of yet, the most important aspect of this new Act is the relaxation of the original timetable for emission control. Congress, however, has kept the 90 per cent reduction as a research goal which will ultimately be met by motor vehicles. Many of the other changes which Congress has instituted in August 1977 concern relatively minor but potentially important housekeeping changes such as:

- changes in the warranty provisions;
- the requirement for inspection and maintenance programmes in problem air quality areas;
- the allowance of wavers of the emission standards for unconventional or new technology engines;
- the allowance for any state to adopt the more stringent California standards as an option to the federal ones.

Throughout this evolution of the United States control programmes, California has always been, by choice, more stringent in its control levels than the rest of the United States. This increased stringency has been justified on the basis of their more severe air quality problem and has resulted in consumers in that state paying a premium for their vehicles because of the small market and the accelerated introduction of technology. From the automotive industry's point of view, it has generally been extremely beneficial to have this two-tiered standard system as it presents a substantial proving ground for their hardware systems to be durability-tested for one to two years before application in the entire United States. A chronology of the emission standards for United States and California cars is tabled in Exhibit 2.4. The most recent Congressional changes and the California emission standards proposals for beyond 1978 model year vehicles are listed in this table. All these emission standards have been converted into an equivalent CVS-CH test procedure for comparability.

Exhibit 2.4.
U.S. NEW AUTOMOBILE EXHAUST EMISSION CONTROL HISTORY

Year	Pollutant Standard (g/km)(1)					
	U.S. Federal (CVS-CH)			California (CVS-CH)		
	CO	HC	NOx	CO	HC	NOx
Uncontrolled	54	5.4	2.5	54	5.4	2.5
1966	NR	NR	NR			
1968	32	3.7	3.1	50%	50%	NR
1970	21	2.4	3.7			
1972	17	1.9	3.1	17	1.9	1.9
1973	17	1.9	1.9	17	1.9	1.9
1974	17	1.9	1.9	17	1.9	1.2
1975	9.3	0.9	1.9	5.6	0.56	1.2
1977	9.3	0.9	1.2	5.6	0.25	0.9
1980	4.4	0.25	1.2	5.6	0.25	0.6(0.9) (2)
1981	2.1	0.25	0.6(0.9)	-	-	-
1982	-	-	-	5.6	0.25	0.25(0.6)

1) Applicable at 80,000 km.
2) Values in parenthesis indicate applicable standard for 100,000 km durability.

2.3. JAPANESE EMISSION STANDARDS

The Japanese emission control programme is a state-controlled programme which was initiated in 1966. The timetable for emission reduction is illustrated in Exhibit 2.5 and detailed discussion of the emission test cycle is in the Appendix.

The Japanese programme has two unique characteristics. First, it imposes a maximum standard for cars and a maximum average level for any test group. It thereby defines the distribution of vehicle emission levels more completely than with the ECE or United States standards. Secondly, it stratifies new motor vehicles into three groups for purposes of standard application:

- new model domestic motor vehicles;
- existing model motor vehicles;
- imported motor vehicles.

The timetable is such that the new model motor vehicles are required to meet the standards before all others then the existing model vehicles are required to meet the standards and finally, the imported vehicles. Generally, there is a one-year delay between each one of these implementation steps. In this manner, the burden of technology development is concentrated on newly designed vehicles.

Exhibit 2.5.

JAPANESE NEW AUTOMOBILE EXHAUST EMISSION CONTROL HISTORY

Year	Pollutant standard(1)			Measurement Unit	Test Cycle
	CO	HC	NOx		
1966	3	-	-	%	4 mode
1969	2.5	-	-	-	
1973	26.0	3.8	3.0	g/km	10 mode
1975	2.7(2.1)(2)	0.39(0.25)	1.6(1.2)	g/km	10 mode
	85.0(60)	9.5(7.0)	11.0(9.0)	g/test	11 mode
1976	2.7(2.1)	0.39(0.25)	1.2(0.85)	g/km	10 mode
	85.0(60)	9.5(7.0)	9.0(7.0)	g/test	11 mode
1978	2.7(2.1)	0.39(0.25)	0.48(0.25)	g/km	10 mode
	85.0(60)	9.5(7.0)	6.0(4.4)	g/test	11 mode

1) Applicable to all new model vehicles that are greater than 1000 kg equivalent inertia weight.

   Later compliance dates are applied to all Japanese manufactured and imported vehicles.

2) The figures show the maximum permissible limits of the standards while the figures in the parenthesis show the average control value.

   All standards applicable at 30,000 km durability and apply to new model vehicles only.

As with the United States programme, the vehicles are required to have pre-production certification of their capabilities. The certification includes a durability run of 30,000 km. In addition, there is an end of assembly line inspection programme which assures that the vehicles are in proper construction when offered for sale to the consumer.

Although the Japanese standard has no maximum idle CO or HC limit for new cars, the extensive in-use inspection programme in Japan limits the idle CO of all in-use cars to 4.5 per cent and the HC to 1,200 ppm. This in-use requirement thus infers a maximum idle limit for new cars.

Only one study on the comparability of the United States and Japanese standards was identified during this study. It was undertaken by the United States EPA(2) and concluded that the 1978 Japanese standards of 2.7/0.39/0.48 (CO/HC/NOx) g/km the 10 mode test and 85/9.5/9.0 (CO/HC/NOx) g/test on the 11 mode test were equivalent to 9.0/0.9/0.6 (CO/HC/NOx) g/km on the United States test. The United States document is very careful to underscore the difficulties in comparing the two test methods. No mention is made in the United States report of the differences in the durability requirement (United States uses 80,000 km while Japan uses 30,000 km). This difference could result in the Japanese standards being effectively more lenient than assessed by EPA although the difference would in all likelihood be slight. In general, very little data is available on the comparison between the two tests and caution should be used in attempting to make exact comparisons.

2.4. ECE EMISSION STANDARDS

The European Community emission standards are recommended by the Economic Commission of Europe (ECE). They first developed standards in 1970 and have followed a control schedule as outlined in Exhibit 2.6. The 1970 ECE standards have been estimated as the equivalent of a 15 per cent control level in Reference 3. The standards are tested according to the ECE driving cycle and a gas analysis procedure in which analysis of total exhaust gas volume is employed. There are two standard levels: prototype and production. Neither of these standards involves any durability testing as is the case with the United States or Japanese standards. In addition to the exhaust emission standards, the ECE also imposes a maximum idle CO limit of 4.5 per cent. This limit must not be exceeded regardless of any engine adjustments which are done on the vehicle with normal mechanic's tools. (The full test cycle is described in more detail in the Appendix.)

The test cycle used by the ECE consists of a modal driving cycle in which the exhaust is measured by means of non-dispersive infra-red (NDIR) for CO and HC and chemiluminescence detector (CLD) for NOx. The test procedures have been recently criticised (4 and 5) in the following problem areas:

- the driving pattern is not representative of current European conditions as determined by the CCMC and thus may not be controlling emissions in critical driving regimes;
- the use of NDIR for HC measurement is not as accurate as the Flame Ionisation Detection (FID) used in the United States and Japan and creates significant error which will become relatively higher as the standards are reduced;
- the current use of full volume exhaust sampling precludes the testing of diesel engines and allows for more condensation of exhaust on the bag walls than does the Constant Volume Sampling (CVS) system used in both the United States and Japan.

As both the United States and the ECE cycles use a form of "cold-soaked" vehicle for the test, a correlation between the two cycles has been partially developed. Work carried out by the

Exhibit 2.6

ECE NEW AUTOMOBILE EXHAUST EMISSION CONTROL HISTORY

Year	Pollutant Standard (g/test)[1]						
	CO		HC		NOx		Idle % CO
	Prototype	Production	Prototype	Production	Prototype	Production	
1970	100-200[2] (31-66)	120-264 (37-79)	8 -12.8 (4.3-6.8)	10.4-16.6 (5.6-8.8)	NR	NR	4.5
1975	80-176 (25-53)	96-211 (30-64)	6.8-10.9 (3.7-5.8)	8.8-14.1 (4.8-7.5)	NR	NR	4.5
1976	80-176 (25-53)	96-211 (30-64)	6.8-10.9 (3.7-5.8)	8.8-14.1 (4.8-7.5)	NR	NR	4.5[3]
1977	80-176 (25-53)	96-211 (30-64)	6.8-10.9 (3.7-5.8)	8.8-14.1 (4.8-7.5)	10 -16[4] (1.9-2.6)	12 -19 (2.2-3.0)	4.5
1979 Proposed	65-143 (20-43)	78-172 (25-52)	6.0- 9.6 (3.2-5.1)	7.8-12.5 (4.2-6.7)	8.5-13.6[5] (1.7-2.3)	10.2-16.3 (1.9-2.7)	4.5

1) No durability testing required.
2) The figures in parenthesis are the equivalent CVS-CH g/km values according to reference.
3) The 4.5 per cent maximum idle was required to be met at any adjustment of the idle system.
4) For automatic transmission equipped vehicles the standards are 12.5-20.0 (15.0-24.0) until 1979.
5) For automatic transmission equipped vehicles the standards are 10.6-17.0 (12.7-20.4) until 1981.

Exhibit 2.7

CORRELATION OF ECE TEST TO CVS-CH
(ECE VEHICLES)

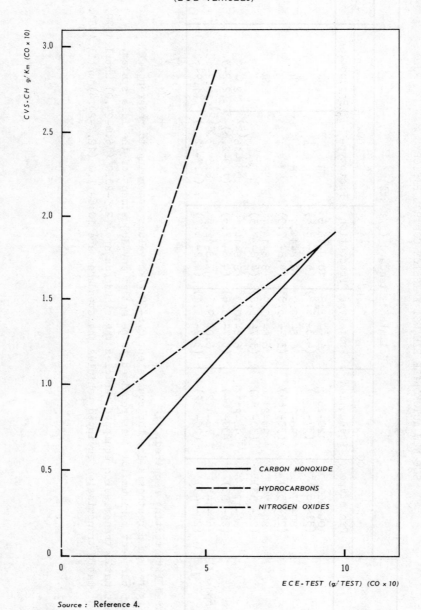

Source: Reference 4.

Umweltbundesamt (UBA) of Germany (4) indicated the linear relationships that appear in Exhibit 2.7. These data were obtained for vehicles which complied with the European test standards and were tested on a CVS-CH test cycle. As the emission control system design is dependent on the standards to which the vehicle must conform, the correlation co-efficients are dependent on the "direction" of conversion (i.e. ECE to CVS-HC vs CVS-HC to ECE). The UBA report showed this clearly in their comparison of United States vehicles run on the ECE cycle to the European vehicle run on a CVS-CH cycle.

2.5. EMISSION STANDARDS IN OTHER COUNTRIES

The current status of automotive emission standards in other major countries is reviewed in Exhibit 2.8. It can be seen that most major countries are currently conforming to the 1973 United States standard level of 17 g/km for carbon monoxide, 1,9 g/km for hydrocarbons and 1.9 g/km for nitrogen oxide. There are a few exceptions where the ECE test cycle has been used as the basis of the standards and in the case of Canada, a slightly more stringent standard than the 1973 United States standard has been applied.

Exhibit 2.8.

CURRENT AUTOMOTIVE EMISSION STANDARDS IN OTHER MAJOR COUNTRIES

Country	Pollutant Standard (g/km)			Test Cycle
	CO	HC	NOx	
Australia	24 (17)	2.1 (1.9)	1.9 (1.9)	CVS-C CVS-CH
Canada	15.5	1.2	1.9	CVS-CH
Hong Kong	100-220 (31-66)	8 -12.8 (4.3- 6.8)	NR	ECE I CVS-CH
New Zealand(1)	100-220 (31-66)	8 -12.8 (4.3- 6.8)	NR	ECE-I CVS-CH
Mexico	24 (17)	2.1 (1.9)	2.2 (2.3)	CVS-C CVS-CH
Sweden	24 (17)	2.1 (1.9)	1.9 (1.9)	CVS-C CVS-CH

1) New Zealand is proposing a dual standard which requires Australian, Japanese and United States manufactured cars to conform to 17/1.9/1.9 g/km standards (United States 73-74).

In reviewing these numbers, one must be extremely careful in assessing the complete control package which is imposed by the country. For instance, in Australia the Federal Government coordinates State bodies in setting vehicle design standards. The States sometimes vary these standards in implementation, to suit local requirements. Thus the current controls nominally require compliance testing at 6,400 km, but to facilitate enforcement, New South Wales has adopted compliance testing at 0 km. Following initial compliance tests, manufacturers can choose to conduct 80,000 km durability tests, or accept a deterioration factor of 1.1. Although in many vehicles 1.1 is an adequate representation of the deterioration factor obtained on a United States type certification run, it may allow some manufacturers to provide vehicles which have less control than the same United States 1973 vehicles. Additionally, this constant deterioration factor method does not, in its present form, make allowance for the effect of fuel composition (lead-free versus leaded) as would be the case with an actual certification mileage accumulation programme. A similar standard has been set in Sweden where they have also adopted a 1.1 standard constant deterioration factor to be applied to the low mileage. Canada has introduced the concept of constant deterioration factors (1.8 for CO and HC and 1.2 for NOx) in a proposed regulation amendment. Also, Canada is at the proposal stage with a maximum idle CO requirement similar to the ECE standard.

3. COST OF EMISSION CONTROL

3.1. INTRODUCTION

3.1.1. Definition of Costs

As mentioned in Section 2.1, the calculation of control costs is a very complex process due to the number of parameters which influence the average cost of emission control. For the purposes of this report, the costs are segmented into three groups:

- initial cost;
- lifetime maintenance cost (10 years/160,000 km);
- fuel consumption cost.

The first two of these items are assessed as normal economic variables and use dollars as the financial measure of impact. Fuel consumption, however, is viewed as a separate commodity for which current prices do not represent true value. The potential undervaluing of the resource has led many governments to assess the impact only in terms of percentage fuel consumption change thus leaving the true economic value undetermined. As the cost of fuel (both present and latent) varies between countries, the sensitivity of fuel consumption changes will also vary. As this report is intended to provide an international perspective of emission control costs, the actual economic cost by country was not calculated. Instead, the impact was assessed in terms of increase (or decrease) in fuel consumption itself.

3.1.2. Definition of Emission Control Level

Before the relationship of cost to level emission reduction can be established, the level of control must be determined. Because of the differences between test method and vehicle durability requirements among countries, the estimation of control level is not a trivial question. As discussed in Chapter 2, the correlation between test methods is known to a degree. However, even after converting the standards to a common test procedure, the standards cannot be directly compared because of the differences in durability requirements.

The ECE standards do not require durability testing but do require prototype certification. The United States and Japanese, on the other hand, are written as standards to be attained at the end of a durability mileage (80,000 and 30,000 km respectively). In order to normalise the standards, the "durability" standards were reduced to account for deterioration and prototype to production slippage. Thus, "effective" design standards were estimated. These values are representative of the level of control that must be attained on the low-mileage prototype in order to pass certification.(5) It is these design restraints which specify the control hardware required and thus the control system cost.

According to EPA certification data,(6) the low-mileage prototype vehicles typically are 60 per cent below the standard for CO and HC and 30 per cent under the standard for NOx. While these factors vary with size of vehicle, technology employed and manufacturer, these averages were used to estimate the low-mileage certification emission rates. In Europe available recent model year data indicate that the manufacturers produce prototypes which have emission rates that are approximately 65 per cent of the standard for all three approval level standards.(7) Thus, this factor was used to estimate the "effective design standards" for ECE vehicles.

Although no comparable data were available from Japan, it is estimated that, because of the durability and end-of-line testing requirements, the correction factor would be the same as in the United States.

3.2. INITIAL COST OF CONTROL

Because the automobile industry has been reacting to other legislation (safety and fuel economy) in addition to the environmental regulations, it is difficult to estimate the cost directly associated with emission control. For instance, although electronic engine control plays a part in emission control, it also is installed to improve fuel economy. Thus, all of its costs should not be solely levied on either programme. As a result of this multiplicity of function, one can create a wide range of cost estimates by simply juggling the cost percentage applicable to a particular programme. This factor is probably the largest source of error between government and industry cost estimates. The divergence of cost estimates is also aggrevated by the adversary relationship between government and industry. This relationship encourages industry to overestimate and government to under-estimate the control cost to support their own vested interests and goals. In analysing the data available, this study has attempted to illustrate the range of costs, and, where possible, the factors that created that range.

3.2.1. United States Estimates of Initial Vehicle Costs

To illustrate the changes that have occurred in cost estimates throughout the last decade, estimates developed by a number of United States studies in the 1972-74 and 1976-77 periods have been compared in Exhibit 3.1. It can be easily seen that the initial estimates in the 1972-73 time frame generally overstated the costs that eventually were incurred. This may have been due to the production cost data being based on automobile industry data or by a lack of analytical sophistication available at that time.

It was not until a 1974 report produced by the National Academy of Sciences (NAS) (11) that a more accurate method of estimation was developed. This improved accuracy can be traced to the use of a sophisticated cost estimation model which "assembles" a vehicle and costs the process according to unit costs which are sensitive to production volume, hardware configuration, material costs and vehicle size. The NAS costs listed in Exhibit 3.1 are based on a 6-cylinder intermediate car which represents the "average" sized car in the United States and thus approximates the average cost. The 1974 NAS costs compare very well with many of the latest cost estimates made in the 1976-77 time frame.

The weight of the vehicle controlled bears a strong impact on costs. Although few studies provide detailed analysis of the vehicle weight effect, a recent EPA document does shed some light on the sensitivity of cost to weight (14). The same study assessed the sensitivity of cost to the design optimisation goal, i.e. cost or fuel economy. The results, illustrated in Exhibits 3.2 and 3.3, indicate that while there is no clear trend towards increasing or decreasing size sensitivity with increasing severity of standards, the fuel optimised vehicles exhibit approximately a 30 per cent cost differential between light (less than 1,400 kg) and the heavy (greater than 1,400 kg) vehicles. This cost differential is not as uniform with the cost optimal assumption which ranges from -9 per cent to 61 per cent.

Predictably, the fuel optimised systems were estimated by EPA to incur more of a cost increase (averaging 80 per cent more) than their cost optimised counterparts. The actual cost of technology will lie between these two design extremes and will be finally determined by the individual manufacturers design optimisation capabilities and the lead-time available. Currently in the United States, as a result of the fuel economy legislation, it is expected that the initial cost of achieving any standard would be close to the fuel optimal cost (i.e. high cost) and slowly would be reduced towards the cost optimal costs (i.e. low cost) because of design improvements.

Exhibit 3.1

U.S. ESTIMATES OF CUMULATIVE INITIAL COSTS(1) OF EMISSION CONTROLS

Emission Standard (g/km)		Estimated Prototype Control Level (g/km)			Estimates Made In 1972-74				Estimates Made In 1976-77			
HC	CO	NOx	HC	CO	NOx	EPA(3)	NAS(4)	In-dustry (5)	NAS(6)	EPA(7)	In-dustry (7)	Gen. Motors (8)	EPA(2,9)
3.7	31.6	(3.1)	1.1	9.5	(2.2)	8	9						
1.9	17.4	(3.1)	0.57	5.2	(2.2)	49	55						
1.9	17.4	1.9	0.57	5.2	1.3	116	76	116	58	33– 98			
0.93	9.3	1.9	0.28	2.8	1.3	249	223	221–284	120	33–129	140		
0.93	9.3	1.2	0.28	2.8	0.84					33–129	140–318	213–250	33–129
0.56	5.6	1.2	0.17	1.7	0.84				178	72–145	174–275		
0.25	5.6	1.2	0.08	1.7	0.84							238–275	68–239
0.25	5.6	0.62	0.08	1.7	0.43							303–360	178–319
0.25	2.1	1.2	0.08	0.63	0.84	375	314	326–438	209	180–221	163–339		68–254
0.25	2.1	0.62	0.08	0.63	0.43					180–225	291–511	353–410	233–389
0.25	2.1	0.25	0.08	0.63	0.17	459	492	305–600	360	285–330	176–637	353–410	285–469

1) All costs in 1977 US dollars per vehicle.
2) The EPA study used the 1977 standards as baseline and then calculated incremental costs. To adjust the estimates to uncontrolled cars the 1976 EPA estimates from reference 12 were used as the baseline values.
3) Cost estimates made in 1972 by EPA in reference 8.
4) Cost estimates made in 1972 by NAS in reference 9.
5) Cost estimates made by industry in 1972, reference 10.
6) Cost estimates made in 1974 by NAS in reference 11.
7) Cost estimates made in 1976 by EPA in reference 12.
8) General Motors published estimate, reference 13.
9) 1977 estimate by EPA in reference 14. Note these costs are sales weighted and the difference represents a cost versus fuel optimised system in 1980.

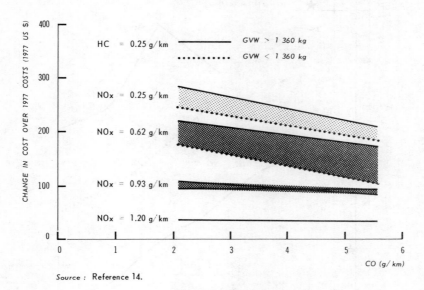

Exhibit 3.2

ESTIMATED CONTROL SYSTEM COST AT VARIOUS STANDARDS
Cost optimal scenario

Source : Reference 14.

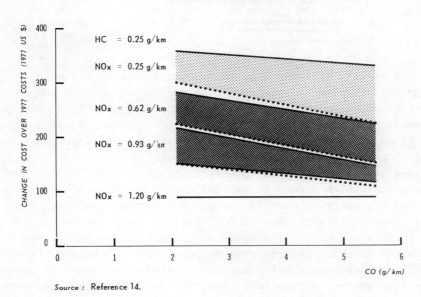

Exhibit 3.3

ESTIMATED CONTROL SYSTEM COST AT VARIOUS STANDARDS
Fuel optimal scenario

Source : Reference 14.

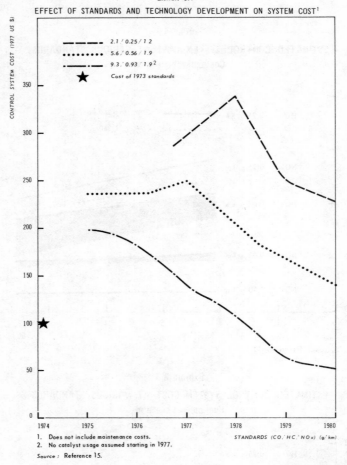

Exhibit 3.4
EFFECT OF STANDARDS AND TECHNOLOGY DEVELOPMENT ON SYSTEM COST[1]

1. Does not include maintenance costs.
2. No catalyst usage assumed starting in 1977.
Source: Reference 15.

The ability of technology improvements to reduce the original design cost is closely related to the designed lead-time and the number of model year production re-design iterations that have been achieved. EPA has estimated the impact of this technology development on costs (15). The relationship determined is illustrated in Exhibit 3.4. The cost improvement per year of any of the standards assessed averaged $40. The data indicate that there is probably a minimum cost which cannot be avoided due to the type of technology used and the material cost involved. For instance, a catalyst would always cost more than an EGR valve not only because of its more intricate construction but also because of the value of material in the device. According to the data provided by EPA, (see Exhibit 3.4) the transient period of cost adjustment is approximately five years. Within this time, the industry would have achieved steady state costs for a given level of control. This cost reduction could result from changing the technology used or cost reductions in the manufacturing of the original control technique.

3.2.2. European Estimates of Initial Vehicle Costs

Although the data sources available from European sources are not as extensive as the United States sources, there have been a number of recent studies on the cost of control as a result of ECE's investigation of further restriction of their standards. This study isolated three major sources of these data:

- reports to ECE by the Bureau Permanent de l'Internationale des Constructeurs d'Automobiles (BPICA) (16, 17);
- the Committee of Common Market Automobile Constructeurs (CCMC) (18);
- the work undertaken by the Umweltbundesamt (UBA) in Germany (19);

Each of these studies will be reviewed in turn using the cost estimates of Industry contained in these studies, since later figures were not available. It has been, however, reported (20) that in the meantime industry estimates have come closer to UBA estimates.

BPICA investigated the cost implications of several standard scenarios. In addition to the average cost, it looked at the impact of the cost by weight size. The general findings of this study group are exhibited in Exhibit 3.5 and indicate that a wide range of costs are expected for all of the standard levels investigated. Because this range was more extreme than that seen in the United States data, Exhibit 3.6 was developed to illustrate the relationships of control level and cost by vehicle weight discretely. The data indicate a high sensitivity of control cost to vehicle weight which is somewhat contradictory to the United States data reviewed in Section 3.2.1. In addition, the average, which in the case of the BPICA study was sales weighted, is substantially higher than those quoted in the United States. This in spite of the fact that the European cars are substantially smaller than the average United States car. According to the correction convention established between the ECE test cycle and the CVS-CH cycle, all the standards investigated by the BPICA Committee were below the levels attained by vehicles produced in the United States up to 1974. The estimates from the United States for the 1973-74 standards ranged from $33 to $98 (1977 US$). According to EPA in its study (21), the lower value should correlate with the small car compliance cost. It would then appear that the BPICA data are very much inflated in relation to the EPA data.

The second data source identified was the CCMC study carried out in conjunction with the ECE standard review. In this study, CCMC looked not only at the cost of attaining a number of control levels but also at the cost dependency of the control technology used. This study expressed the cost increase in terms of percentage

Exhibit 3.5

BPICA (ECE) COST DATA (1,2)

Standard (CVS-CH)[3]			Cost (Jan. 1977 US$) per vehicle
CO (g/km)	HC (g/km)	NOx (g/km)	
31-66 (38)[4]	4.3-6.8 (5.4)	1.9-2.6 (2.4)	Baseline
20-43 (25)	3.2-5.1 (4.0)	1.9-2.6 (2.4)	69-260 (77)
20-43 (25)	2.8-4.4 (3.5)	1.9-2.6 (2.4)	85-428 (130)
20-43 (25)	2.8-4.4 (3.5)	1.7-2.3 (2.1)	88-316 (105)
19-40 (24)	2.6-4.1 (3.3)	1.9-2.6 (2.4)	111-558 (215)
17-36 (21)	2.4-3.7 (3.0)	1.9-2.6 (2.4)	162-782 (284)
15.5-33 (19)	2.2-3.4 (2.8)	1.9-2.6 (2.4)	180-1286 (339)
15.5-33 (19)	2.8-4.4 (3.5)	1.9-2.6 (2.4)	NA (282)

1) Data extracted from reference 16 and 17.
2) Estimated cost above the 1971 ECE 15 standards.
3) Original standards were converted to CVS-CH equivalent using data from reference 4.
4) The values in parenthesis are the sales weighted averages based on the 1975 market by inertia weight.

NA. Not available.

of the total vehicle cost. In order that a quantifiable dollar amount could be produced, the average new car price developed in the BPICA study was applied to these percentage values.(*) The estimates which it developed for a variety of control hardware systems are listed in Exhibit 3.7. This same exhibit lists the EPA estimates for comparable systems. It is immediately apparent that similar to the BPICA study, there are large discrepancies in these prices. The current hardware systems, that is the engine modifications, are probably costed correctly by CCMC as they are below the EPA value which, it must be remembered, is based on a larger average sized vehicle. The "advanced technology" systems, however, average 70 per cent above their comparable EPA cost estimates. This difference is compounded when the size difference between the average ECE and the average EPA cars is taken into account.

---

*) The BPICA study estimated the May 1976 average car price in Geneva to be US$5,500 (1977 prices).

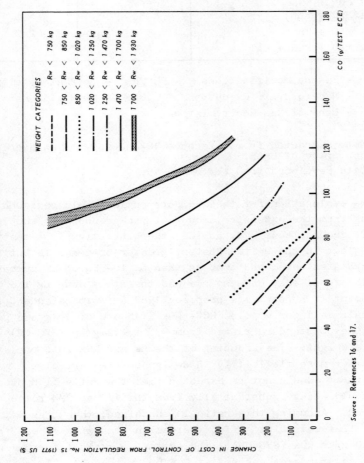

Exhibit 3.6

BPICA ESTIMATES OF COSTS OF ATTAINING VARIOUS CONTROL LEVELS BY VEHICLE WEIGHT

Source: References 16 and 17.

Exhibit 3.7
COMPARISON OF CCMC AND EPA ESTIMATED COST OF CONTROL HARDWARE

System Hardware	Average Reported Cost per vehicle (1977 US$)	
	CCMC[1]	EPA[2]
EM	55	75
EM + MAO	220	130
EM + MAO + EGR	330	160
EM + MAO + TR	370	240
EM + MAO + EGR + CAT	470	260

EM   - Engine Modifications   TR - Thermal Reactor
MAO - Manifold Air Oxidation   CAT - Oxidizing Catalyst
EGR - Exhaust Gas Recirculation

1) Data extracted from reference 18 and assumes an average new car price of US$5,500.
2) Data developed from reference 15.

No explanation for these discrepancies could be found, however, judging from the extensive amount of background data supplied with the EPA documents as well as the other investigations that have been carried out in the United States, greater credence is put on the United States developed numbers than on the European numbers.

The final cost data source was the study done by the German government to support its proposed 1982 emission standards. These proposals call for a joint HC + NOx standard of 10 grams per test while the CO standard ranged from 7.5 to 12 g/km (CVS-CH) based on inertia weight. The grouping of the HC and NOx into one standard allows for more flexibility in design on the part of the manufacturer, however, it should not be expected that the ratio of hydrocarbon to NOx should change substantially from the 1/3 to 2/3 ratio currently attained. Assuming this ratio is sustained, then these standards can be translated into effective design standards of 0.65 g/km HC- and 1.0 g/km NOx (CVS-CH). The estimated costs incurred in attaining these standards was set at US$160 by the UBA. If these standards are related back to the United States prototype control levels (see Exhibit 3.1) it is apparent that these standards are comparable - as far as CO and HC emissions are concerned - with the United States 1975-77 standards and - as far as the NOx emissions are concerned with the 1973 United States standards. The UBA estimates are thus very similar to the United States estimates. It should, however, be mentioned that the UBA estimates are not limited to the reduction of exhaust emissions but include efforts to reduce at the same time the fuel consumption, although already lower than in the United States.

3.2.3. Japanese Cost Estimates

Only a limited amount of Japanese emission control cost data was available and only represented government estimates (52, 53, 54, 55). The estimated increase in the vehicle costs between the 1976 and 1978 standards is $50-330 (based on a 300 Yen/US$ exchange rate). This is an extremely wide range and was based on the assumption that the entire 10 per cent increase in cost between the 1976 and 1978 model year vehicle was due to emission controls. This may overestimate the cost impact of controls as there were concurrent increases in labour, material and energy costs during the same period. It is, however, similar in magnitude to the United States estimates of cost increases resulting from the imposition of 1980 standards (see Exhibit 3.2 and 3.3).

3.2.4. Other Cost Estimates

After this review of both United States and European cost estimates, one is left with a dichotomy of cost estimates. Although "third party" cost estimates are limited, data provided by the Australian Government on its estimated cost for its ADR 27A regulations, which are copies of the 1973 United States standard, indicate that the expected range of cost over an uncontrolled vehicle is US$60 to US$216 with an average of US$120. This cost is probably high compared to European and United States costs for the same control due to the low sales volumes of vehicles (by model) in Australia. Because of the impressive amount of documentation and industry concurrence in general with the EPA cost estimates, this study used the United States values for the representative cost of control.

3.3. EFFECT OF EMISSION CONTROL ON FUEL ECONOMY

As noted previously in Section 3.1, fuel economy impact is assessed in this report separately from the cost impacts of emission controls. As with the costs, the impact will be assessed on a per car basis. As the majority of the data on fuel economy are of United States origin, the analysis presented in this report is biased toward the American experience. To compensate for this, wherever possible, the data from the United States have been divided into light and heavy weight of vehicles to indicate the impact of regulations on small sized car fleets.

The sensitivity of fuel economy to emission control is related to a number of factors of engine design and control. Exhibit 3.8 gives the general effects these factors have on fuel economy and exhaust emissions. Because of the range of effects that these factors have, it is not merely enough to know the direction one is going in the control of exhaust emissions to deduce the direction change in fuel economy. The National Academy of Sciences coined the phrase "decoupling" to refer to certain devices which have no effect on fuel economy while containing emissions. These devices which are generally after-treatment devices, allow the engine to be calibrated for optimal fuel economy and leave the reduction of exhaust concentrations to the after-treatment device. As a result of these devices, EPA has concluded that "At a fixed emission level fuel economy is a function of the usage of fuel efficient control technology".(22)

Exhibit 3.8
IMPACT OF VARIOUS EMISSION CONTROL TECHNIQUES ON FUEL ECONOMY

Control Technique	Pollutants Controlled	Fuel Economy Effect
Retarded Spark Timing	HC, NOx	Negative
Rich air/fuel ratio	NOx	Negative
Lean air/fuel ratio	HC, CO, NOx	Positive
Port EGR	NOx	Negative
Proportional EGR	NOx	None or Positive
Quick Heat Intake Manifold w/fast choke	HC, CO	Positive
Heated intake air	HC, CO	Positive
Air injection	HC, CO	Almost none
Oxidation catalyst	HC, CO	None
Reduction catalyst	NOx	None
Thermal reactor	HC, CO	None
Reduced compression ratio	HC, NOx	Negative

Source: Reference 22.

In Exhibit 3.9, the relationship of percentage change and fuel economy from a 1957/67 base average to inertia weight is portrayed. These data were extracted from EPA certification data (23). It can be seen that with the imposition of the 1973-74 standards there was no negative effect on the vehicles below 1,600 kg in inertia weight. In fact, these vehicles showed a 7 per cent improvement from the base case. The larger models, however, showed significant effects on fuel economy due to emission control, i.e. an average 16 per cent reduction in their fuel economies. EPA attributes this loss in fuel economy solely to the selection of control technology. In 1975 models, all vehicles regardless of inertia weight improved their fuel economies with the large vehicles showing a net zero effect of the emission control imposition. This improvement occured in spite

Exhibit 3.9
PERCENT CHANGE IN FUEL ECONOMY FROM
1957-1967 AVERAGE Vs INERTIA WEIGHT

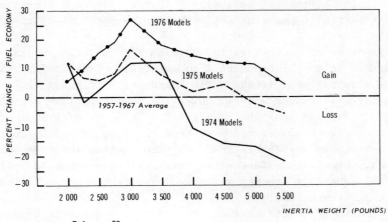

Source: Reference 23.

Exhibit 3.10
EFFECT OF STANDARDS AND TECHNOLOGY
DEVELOPMENT ON FUEL ECONOMY

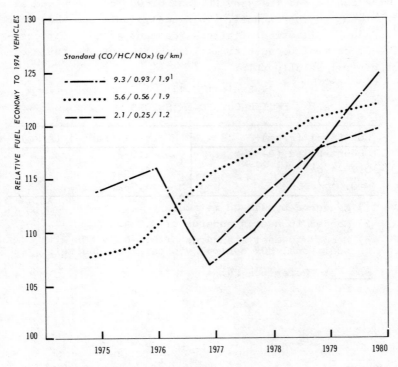

1. No catalyst usage assumed starting 1977.

Source: Reference 22.

of the fact that the 1975 models were more stringently controlled than the 1974 models. The 1976 models, which were at the same control level as in 1975, continued this improvement trend with the small cars indicating approximately a 15 per cent improvement and the large cars a 11 per cent improvement from the 1957-67 base average.

This trend of technology to optimise for fuel economy was analysed further in more recent EPA documents(24). Exhibit 3.10 illustrates the improvement trends which EPA expects to see in fuel economy at any one standard level. Over the 5-year period for 1975 to 1980, it expects that the relative fuel economy of vehicles at all the control levels assessed to rise from approximately 10 to 20 per cent above the levels achieved when meeting the 1974 emission standards. EPA also believes that this fuel economy would still allow for technology changes which would allow the optimisation of the costing of vehicles.

The basic trade-off that occurs for fuel economy is between NOx and HC control. Exhibit 3.11 tables EPA's estimates of the "engine-out" HC emissions at various NOx control levels that are required to sustain good fuel economy. In order to create this table, EPA has assumed that there was no hydrocarbon emission control on the engine and that any HC control would take place as an after-treatment to the exhaust. This analysis is somewhat biased towards EPA's predilection of after-treatment as the most cost-effective of control because of its decoupling of the fuel economy emission control relationship.

Exhibit 3.11
ENGINE-OUT HC EMISSIONS [1]

NOx Level (g/km)	HC Level (range) (g/km)
Uncontrolled (2-3)	1.0-2.0
0.75-1.0 [2]	1.1-2.4
≈ 0.6 [3]	1.2-2.6

1) Measured by CVS-CH method.
2) Needed to meet standards of 0.9-1.2 NOx.
3) Needed to meet standards of 0.6-0.25 NOx with additional NOx control via catalyst control.

Source: Reference 24.

Summarising the United States experience to the present, the United States has found that there is effectively no fuel economy penalty up to and including the 1977 United States standards. However, there has and will continue to be transient fuel economy penalties as the industry learns to optimise its vehicle designs for both cost and fuel economy. EPA, in fact, is currently estimating that there will be little or no fuel economy loss at standard levels as stringent as 0.25 HC, 2.1 CO, 0.62 NOx (g/km) provided adequate lead-time for development and optimisation is allowed(25).

The general findings by EPA are in conflict with the conclusions of a study undertaken for the Economic Commission for Europe by the Bureau Permanent de l'International de Constructeurs d'Automobiles (BPICA)(26). In its investigation of relationship between emission limits and fuel economy, BPICA estimated that there would be between a 1.2 and 3.3 per cent increase in fuel consumption over the range of standard scenarios that it investigated (see Exhibit 3.12). Emission control levels which the ECE study group investigated are approximately those achieved by the American programme in the 1970 through 1973 period. Although North American cars, at that time, experienced up to a 10 per cent loss in fuel economy this loss, according to EPA's analysis, was due entirely to the non-optimal control systems used on those vehicles. In fact, according to Exhibit 3.9 the small cars which met these standards achieved an improvement in fuel economy (approximately 7 per cent). The BPICA study team may, in fact, have produced estimates which represent the cost optimal vehicle design which would create a fuel penalty associated with standards. Thus, it would seem that they have stated the highest fuel consumption impact possible.

The Japanese data (52) on the vehicles designed to achieve their 0.25 g/km (1978) NOx standards indicate that the vehicles are capable of attaining fuel economies as good as the 1976 vehicles and in some cases better than the 1973 vehicles. Exhibit 3.13 presents data on some of the better fuel economy vehicles from Japan. The Japanese experience appears to be very similar to that of the United States in that they are not expecting any fuel economy impacts due to the imposition of their latest standards which are approximately equivalent to 1981 United States NOx standards.

In summary, it would appear that up to control levels approaching the United States 1980 standards, there should be no net fuel consumption penalty because of the application of control standards. In any jurisdiction imposing standards there may, however, be transient fuel economy penalties due to the non-optimal design of the control system. These transients should last for 1 to 5 years at which time any penalty effect would be eliminated.

Exhibit 3.12

PERCENTAGE INCREASE IN FUEL CONSUMPTION IN ECE AUTOMOBILES
AS A RESULT OF REDUCED EMISSION STANDARDS(1)

Certification Standards (g/test)			Effective Design Standards (g/km)(2)			Fuel consumption increase(%) Equivalent Inertia Weight (kg)		
CO	HC	NOx	CO	HC	NOx	<910	910-1,130	>1,130
65-145 (21- 44)	6.0-9.6 (3.3-5.1)	10 -16 (1.9 -2.6)	14-29	2.1-3.3	1.2-1.7	1.3	1.1	≤ 1.3
65-145 (21- 44)	6.0-9.6 (3.3-5.1)	8.5-13.6 (1.7- 2.3)	14-29	2.1-3.3	1.1-1.5	3.1	3.2	3.8
65-145 (21-44)	5.2-8.3 (2.8-4.4)	10 -16 (1.9- 2.6)	14-29	1.8-2.9	1.2-1.7	2.7	1.7	2.9

1) Values in parenthesis represent equivalent g/km on CVS-CH cycle.
2) CVS-CH values adjusted as per section 3.2.

Source: Reference 26.

Exhibit 3.13

EFFECT OF JAPANESE EMISSION CONTROL STANDARDS ON FUEL ECONOMY (km/l)

Engine Displacement (cc.)	FY1976 Standards			FY1978 Standards		
	Min.	Max.	Avg.	Min.	Max.	Avg.
Over 2,000	4.8	7.4	6.4	5.1	8.2	7.3
About 2,000	6.5	10.0	8.1	7.8	11.5	9.6
About 1,800	7.6	12.5	10.3	10.0	13.0	11.6
About 1,600	7.0	12.0	9.9	9.3	14.0	12.1
About 1,400	11.0	13.0	12.1	10.5	15.5	13.6
About 1,200	10.0	14.5	12.1	14.0	17.0	15.1
About 1,000	-	-	-	-	-	19.0
About 1,000	14.5	19.0	16.8	15.5	21	16.0
(Rotary engine)	6.2	6.7	6.5	-	-	6.5

Note: 10 mode data.

3.4. MAINTENANCE COST OF EMISSION CONTROL

There have been relatively few studies on the maintenance cost increases associated with emission controls. What studies have been completed have generally attempted to estimate the increase in cost due to the maintenance for emission controls. As with initial price, the maintenance items which are included can vary depending on the analysts perspective, e.g. spark plug replacement could be included even though this was a requirement before the advent of emission controls. The data collected has all been adjusted to reflect the estimated per vehicle costs over 10 years or 160,000 km.

As many of the studies calculate the cost assuming all required control system maintenance is done, the cost estimates will be higher than what is actually incurred by the consumer. This is particularly true when the cost of catalyst replacement is added to the lifetime costs as it is extremely unlikely this event will occur given the present regulatory system.

The maintenance costs are very sensitive to the amount of lead in the gasoline. Although the addition of lead has beneficial effects as far as fuel economy and vehicle performance are concerned it and the chemical scavengers with the lead act as contaminants and corrosive agents in the engine and exhaust system.(27) This, while the United States studies (Exhibit 3.14) indicate a zero or negative cost with advanced controls, this is in large part due to the leadfree fuel used.

The cost tabled in Exhibit 3.14, do indicate some consensus among the governments on the maintenance costs. Of particular interest is the United States data which indicates low cost

increases until the most stringent United States goals which would
jump the costs by $90-190 depending on the data source. It should
be noted that these "ultimate system" costs are probably high due
to the experimental nature of the proposed control systems required
to meet the standards. Of equal importance is the fact that both
the EPA and the NAS estimates for the advanced catalyst technology
systems base their cost on the assumption that a catalyst change
would occur at 80,000 km. The likelihood of this catalyst replacement is severely in doubt given the present field enforcement programme in the United States. If there is no replacement, the lifetime costs should decrease by approximately $135 leaving the cost
estimate in the order of $140 or $14 per year. For the most stringent United States emission goal, which would result in approximately
$400 to $460 with a catalyst change, the non-replacement of catalyst
would lower the cost of between $30 and $90 ($3 to $9 per year).
It should be noted that the uncertainty of the costs is much higher
with the long-term emission goal limits.

Exhibit 3.14
ESTIMATES OF LIFETIME CONTROL MAINTENANCE COSTS

Emission Standard (g/km)			Effective Design Standard (g/km)			Estimated Lifetime Control Maintenance Costs (1977 US$)			
HC	CO	NOx	HC	CO	NOx	EPA (1)	NAS (2)	UBA (3)	Ford Europe (4)
3.7	31.6	(3.1)	1.1	9.5	(2.2)	224	0		472
1.9	17.4	(3.1)	0.57	5.2	(2.2)	224	327		
1.9	17.4	1.9	0.57	5.2	1.3	224	387	90-320	
0.93	9.3	1.9	0.28	2.8	1.3	252	255(5)		
0.93	9.3	1.2	0.28	2.8	0.84				
0.56	5.6	1.2	0.17	1.7	0.84		270(5)		
0.25	5.6	1.2	0.08	1.7	0.84				
0.25	5.6	0.62	0.08	1.7	0.43				
0.25	2.1	1.2	0.08	0.63	0.84	280	270(5)		
0.25	2.1	0.62	0.08	0.63	0.43				
0.25	2.1	0.25	0.08	0.63	0.17	392	460(6)		

1) Reference 28.
2) Reference 29.
3) Reference 30.
4) Reference 31.
5) Assumes a catalyst change at 80,000 km costing $135.
6) Assumes a catalyst change at 80,000 km costing $371.

The German data from UBA (30) indicates good correlation to the United States data in the same control regime. They estimate that their newest 1980 standard proposals would increase lifetime maintenance costs by $90-320.

The only other European estimate was obtained from Ford of Europe.(31) They estimate that the lifetime cost for the current European standards is $472. This is significantly above any other estimates. As each of these studies assumes different definitions of marginal maintenance it is unclear as to the reasons for this discrepancy.

The Japanese data available (52) estimates a $10-20 annual maintenance cost increase will be imposed by the 1978 standards. This results in a $100-200 lifetime maintenance cost which is slightly lower than the United States estimates.

4. PERFORMANCE OF EMISSION CONTROL SYSTEMS IN USE

4.1. INTRODUCTION

Although the cost of emission control systems is a direct function of the original design limits, the effectiveness of emission controls is measured by the final emission rate which is achieved when the vehicle is used by the consumer. Because of the durability requirements in both the United States and Japan for certification of new production vehicles, there is sometimes a misunderstanding of the meaning of the deterioration factors which are a product of the certification testing. The factors obtained during the certification process represent the best possible performance of the control system. In the case of the United States factors, the manufacturer is responsible for the mileage accumulation and can perform certain amounts of maintenance during the 80,000 km. This maintenance is carried out by the best mechanics available to the company and, in addition, the vehicles that are used are hand-built models which will, in all likelihood, have superior components and assembly than the production models. Thus, the deterioration rates bear little resemblance to actual field performance of the control systems.

In order to distinguish between the "official" deterioration factors and the general decline of the control system's effectiveness in the field, the term degradation will be used to indicate field performance abilities.

The expense and complexity of testing vehicles on the full test cycle, has resulted in relatively little field assessment work being carried out. The United States has probably been the most active and has had several programmes on-going throughout the last few years. The most significant of these is the emission factor development programme (32) which produces data to be used in a handbook for emission inventory calculation.(33) California also has carried out several programmes to assess the performance of emission controls in consumers' hands. (34, 35)

Only the control effectiveness, as measured by the field testing of the vehicle, will be assessed in the following Sections 4.2 and 4.3. The reasons for these large increases in emission rates will be discussed and analysed in Section 4.4. Alternative control strategies will be assessed based on these field performance data in Section 4.5.

4.2. CURRENT CONTROL SYSTEMS DEGRADATION

Based on the information gathered in a variety of field surveillance programmes, EPA has developed emission factors which represent the average emission rates of vehicles by control type and age. For the purposes of this study, the assumption will be made that the average emission rate of a vehicle over its lifetime will be the same as the emission rate in its fifth year of operation or at approximately 80,000 km. Using this assumption, Exhibit 4.1 was developed from data presented in the most recent addition of EPA's Mobile Source Emission Factors(33). Exhibit 4.1 clearly indicates that the emission rates of vehicles in use are grossly above the originally mandated standard at 80,000 km. With the exception of the 1970 (3.7/31.6) vehicles, the HC averages 1.75 of the standard at 80,000 km while the CO for the most recent levels of emission control is 2.35 of the standard. The NOx which was controlled only after 1973, indicates a much closer compliance with the standard for the 1973 (1.9/17.4/1.9) and the 1975-76 (0.93/9.3/1.9) standards. However, EPA expects that the latest 1977-78 standards (0.93/9.3/1.2) will be some 60 per cent above the mandated standard at 80,000 km usage.

These general conclusions are supported by other investigators' work as illustrated in Exhibit 4.2. The low mileage Swedish data support the industry's contention that they are producing cars which meet the required standard in Sweden (the 1973 United States standards), however, the limited Swedish testing programme on high mileage vehicles indicates a substantial deterioration to levels which could be construed as close to uncontrolled. The Swedish document from which these data were extracted states that "the results obtained so far seem to show that emissions from the average old vehicles are much higher than were recorded for vehicles of the same models when almost new". A similar pattern was also seen in surveys carried out by the California Air Resources Board (34) and the New York State Department of Environmental Conservation(37). In general, all of these researchers have qualified their damnation of the emission control durability by stating that the control systems themselves appear to have the capability when properly maintained of meeting the emission standards. This subject will be covered further in Section 4.5.

Exhibit 4.1.

EPA ESTIMATED AVERAGE IN-USE EMISSION RATES FOR CURRENT CONTROL SYSTEMS

Original Standard (g/km)			In-Use Emission Rate(1) (g/km)			In-Use Rate/Standard		
HC	CO	NOx	HC	CO	NOx	HC	CO	NOx
Uncontrolled			5.0	54	2.1	-	-	-
3.7	31.6	-	3.4	40	2.9	0.92	1.26	-
1.9	17.4	-	3.4	40	2.9	1.79	2.30	-
1.9	17.4	1.9	3.4	40	2.1	1.79	2.30	1.11
0.93	9.3	1.9	1.6	22	2.0	1.72	2.37	1.05
0.93	9.3	1.2	1.6	22	1.9	1.72	2.37	1.58

1) Average emission rates in fifth year of operation as measured by CVS-CH method.

<u>Source</u>: Reference 33.

Exhibit 4.2.

ESTIMATES OF THE FIELD PERFORMANCE OF EMISSION CONTROLS

Data Source	Original Standard (g/km)			Estimated In-Use Emission Rate(1) (g/km)		
	HC	CO	NOx	HC	CO	NOx
Sweden(2)	1.9	17.4	1.9	1.9	14	1.5
Sweden(3)	1.9	17.4	1.9	4.6	69	-
CARB(4)	0.56	5.6	1.2	0.84	9.3	1.7
NY State(5)	0.93	9.3	1.9	1.3	20	2.2

1) All data at 80,000 km.
2) Data developed from grouped data averages of tests with low mileage ($\simeq$ 5,500 km) vehicles presented in reference 36.
3) Based on limited high mileage (72,000 km) tests.
4) California Air Resources Board - reference 34.
5) New York State Department of Environmental Conservation - reference 37.

4.3. ADVANCED CONTROL SYSTEM DURABILITY

The question that immediately arises after review of the current durability of the control technology is what improvements are likely with the use of advanced technologies. Assuming for the moment that no improvements are made in the adjustability limits of the vehicle (this will be discussed in Section 4.5), the durability is closely related to the type of technology employed for the control system. While everyone can agree that catalyst systems are very sensitive to lead, it is also true that a non-catalyst system is sensitive to lead and will exhibit higher degradation factors because of the use of a leaded gasoline. However, with the exception of a diesel engine, there should be very little difference in the degradation characteristics (as measured as a ratio of standard at 80,000 km) between various engine control types. This will be particularly true with the introduction of closed feedback control systems such as would be used with the three-way catalyst systems. In the extreme, these systems, whatever the control technology used, will be able to adjust their operating conditions to maximise fuel economy and minimise emissions.

EPA, in their emission factors development programme, have estimated that the expected 80,000 km in-use emission rate of a 0.25 g/km HC vehicle would be 1 g/km. Likewise, the CO and NOx show marked increases (refer to Exhibit 4.3.). Thus, EPA is currently estimating that there will be no improvement in the degradation ratio of the advanced control systems. In fact, it forecasts all systems to become worse and, in the case of the 0.25 g/km NOx standard, EPA admits that it is probably underestimating the degradation factor of this vehicle.

4.4. REASONS FOR HIGH DEGRADATION RATES

In the previous sections, it has been shown that the control systems that are installed by the manufacturer just simply do not have the durability in the field. Thus, the effectiveness of the control programme is greatly reduced. In order to ascertain the reason for these high degradation rates, several field surveillance programmes have been completed primarily by the United States.

Most recently, California has been extremely active in this area. Their most recent surveillance report of 1975 and 1976 California (38) specification vehicles, shown, in Exhibit 4.4, the incidence of component maladjustment or failure in these vehicles. This programme was carried out in the 1976/77 time period when the vehicles tested had relatively low mileage on them. Even so, the

Exhibit 4.3
EPA ESTIMATED AVERAGE IN-USE EMISSION RATES
FOR ADVANCED CONTROL SYSTEMS

Pollutant	Standard (g/km)	Emission Rate at Age y(1)(g/km)	Emission Rate In 5th Year	Emission/ Standard Ratio
HC	0.25	0.12 + 0.16y	0.9	3.7
CO	5.6	3.8 + 1.82y	12.9	2.3
CO	2.1	1.4 + 1.22y	7.5	3.6
NOx	1.2	0.9 + 0.14y	1.6	1.3
NOx	0.62(2)	0.4 + 0.17y	1.3	2.0
NOx	0.25(3)	0.1 + 0.08y	0.5	2.0

1) Where y is the age of the vehicle.
2) At this level NOx is assumed to be catalytically controlled.
3) Because of a lack of data on the deterioration of 0.25 NOx vehicles the per cent deterioration rate is assumed to equal that of the 0.62 NOx vehicle. This may understate the deterioration at that level.

Source: Reference 33.

data indicate that 44 per cent of the vehicles failed to meet the emission standards for which they were designed. The most frequent pollutant failure was CO. This judgement has been shown both in the California surveys and EPA's work(39). Exhibit 4.5 illustrates the now typical distribution of emission failures. The causes for the failures are numerous but the most frequent is carburetor idle mixture maladjustment. The maladjustment of the carburetor idle mixture has been achieved by the service industry by removal of limiter caps.(39) Their estimates of carburetor maladjustment are tabled in Exhibit 4.6. The occurence of a relatively high frequency of engine idle speed maladjustment can be directly related back to the idle CO circuit also. There are, of course, the odd vehicles which would have proper idle CO yet high idle speed but these are relatively rare and for the most part the two failures should be considered as one problem, in proper idle mixture control.

It has also been shown in Great Britain (40) in tests on vehicles in use that similar problems occurred on the less controlled British vehicles. On average, the error on static timing was 2.6° retarded while the CO percentage was 2.6 per cent above the correct setting. Idle speed was also noted in these reports as commonly incorrect.

Of the other failure modes, a more detailed examination of the failure frequency by component is listed in Exhibit 4.7. These data

Exhibit 4.4

INCIDENCE OF COMPONENT MALADJUSTMENT OR FAILURE
IN 1975/76 CALIFORNIA VEHICLES

(Percentage)

Problem	All Vehicles	Passed CVS-75	Failed(1) CVS-75
Carburetor Idle Mixture	42	11	31
Engine Idle Speed	23	11	12
Ignition Timing	18	5	13
Failed/Clogged Components	19	2	17
Disconnected Systems	9	3	6
Tampering(2)	15	1	14
Any of above	66	22	44
None	34	34	0
Total	100	56	44

1) Emission(s) more than 15 per cent over applicable standards.
2) Tampering is defined as a modification which clearly shows intent to defeat emission controls (e.g., a steel ball in a vacuum line, a hole punched into a vacuum diaphragm, air injection pump disconnected, etc.).

Source: Reference 38.

were developed by Champion Spark Plug Company in a large field survey undertaken by them in recent years. The fact that air filters are commonly outside of specification on vehicles will compound the effect of that idle CO mixture while the high incidence of spark plug malfunction and in general the ignition system failure rates will cause an elevation of the hydrocarbon level.

The other important question which must be answered concerning the degradation of these control systems is relationship between vehicle malfunctions or maladjustments and vehicle use or age. Data on idle test programmes tend to indicate that the degradation of the idle CO occurs early in the vehicle's life. To exemplify this situation which has been documented by many investigators (41, 42, 43), results from a set of idle surveys carried out in Ontario, Canada are presented in Exhibit 4.8. From these graphs, it is easily seen that the majority of the degradation occurs within the first 20,000 miles or the first two years of the vehicle's life at which time the vehicle attains a form of "steady state" level of idle CO. The degradation of idle hydrocarbons (see Exhibit 4.9) with vehicle use is not so rapid and generally has been shown to exhibit a more constant deterioration rate. This characteristic is predictable as a

Exhibit 4.5

NUMBER OF 1975 VEHICLES THAT FAIL THE FEDERAL STANDARDS, BY STANDARD FOR THE CITIES OF CHICAGO, HOUSTON, ST. LOUIS, WASHINGTON, AND PHOENIX

Manufacturer	Total Tested	Total Number Failures	Per cent Failure	Number of Vehicles That Fail						
				Only HC	Only CO	Only NOx	Both HC, CO	Both HC, NOx	Both CO, NOx	All Three Standards
GM	229	139	61	3	41	24	53	1	8	9
Ford	124	75	60	2	30	16	9	4	8	6
Chrysler	77	68	88	0	24	4	31	0	4	4
AMC	23	16	70	2	5	6	2	1	0	0
U.S. Total	453	298	66	7	101	50	95	6	20	19
Foreign Total	134	73	54	8	24	8	17	5	1	10
Overall Total	587	371	63	15	125	58	112	11	21	29

Source: Reference 39.

Exhibit 4.6

FREQUENCY OF IDLE CO MALADJUSTMENT (PERCENT) FOR 1975 VEHICLES

Data Source	Sampling Size	Frequency of Idle Maladjustment From Spec. (%)	Idle Limiter Caps Removed (%)
Champion Spark Plug Company	298(1)	38.9	Not reported
EPA Emission Factors Programme	Survey 1 364(2)	28.3	Not reported
	Survey 2 115	30.4	34
Mobile Sources Enforcement Division	190	Cannot be determined	26.5

1) Specification was idle CO greater than 2.0 per cent.
2) Specification was idle CO greater than 1.5 per cent.

Source: Reference 39.

Exhibit 4.7

FREQUENCY OF FAILURE FOR SELECTED COMPONENTS (ALL MODEL YEARS)

System/Component	Sample Size	Freq. of Failure(%)
Points	3877	21.0
Cap & Rotor	4625	5.1
Ignition leads	4626	18.7
Spark Plugs	4626	31.7
PCV	4624	4.6
EGR	1247	4.0
Belts (Replaced)	4609	14.0
Hoses	4528	10.8
Air Filter	4369	34.5

Source: Reference 39.

result of the full component failure that occurs in the ignition system or through gradual wear of the cylinder wall, etc. Again, the Canadian survey data are used to exemplify the type of data that have been collected from surveys carried out in the United States, New Zealand, Australia, Japan and Europe.

The figures 4.8 and 4.9 also indicate that the most significant determinant of exhaust emissions (per cent CO) is the model year of vehicle as the model year defines the level and type of emission control technology employed on the vehicle. The ultimate improvement attained by an inspection/maintenance system is, therefore, limited by original design. Thus, although a very effective inspection system could reduce the per cent CO of 1970 cars to 3.3 per cent from 4 per cent (Graph 4.8A), the inspection system could never attain the lower levels of the 1975 cars. Therefore, inspection systems will never fully replace the benefit of more stringent design standards.

The problem of durability then appears to be related to the ability of the consumer or the service industry to alter the vehicle from the manufacturer's specifications. Exhibit 4.10 indicates the generalised relationship of idle CO to carburator setting. For a typical currently available vehicle, there is a very high CO per cent sensitivity at the point where the vehicle must be set and thus the chance of error is very high. Also, on many cars there is no effective upper limit to the CO per cent thus allowing 10 per cent CO setting to be attained. Although with the ignition system there are some adjustments which can be altered such as basic timing, it appears from the data reviewed that it is more of a component failure problem than a maladjustment problem. For the NOx control, however, it seems to be a tampering problem as opposed to a maladjustment or a poor servicing of the system. The exhaust gas recirculation (EGR) is additionally susceptible to high degradation because of clogging of the orifies caused by the use of high lead content fuels.

One other question that is associated with the degradation is: "Do people in fact service their vehicles according to the recommended maintenance schedule"? There have been several surveys in the last few years which have investigated this question. The most recent is from California surveillance programme, (44) the results of which are shown in Exhibit 4.11. They indicate that, at least for the low mileage vehicles, servicing is attempted by the consumer at levels which approach the manufacturer's recommended servicing frequency. While the correlation between the level of maintenance and emissions is present for CO it is not enough of a factor to account for the high deterioration that does take place (see Exhibit 4.12).

The emission problem seems to reduce to three major factors which contribute to the lack of emission control by vehicles in consumers' hands:

  i) The ability of the control systems to be adjusted;
 ii) The service industry's propensity to either not identify engine problems or to adjust the engine outside of manufacturer's specifications;
iii) The low incentive for the service industry to maintain control systems coupled with the high incentive to satisfy

Exhibit 4.8 A
DEGRADATION OF IDLE CARBON MONOXIDE WITH VEHICLE USE
Engine size 50-140 c.i.d.

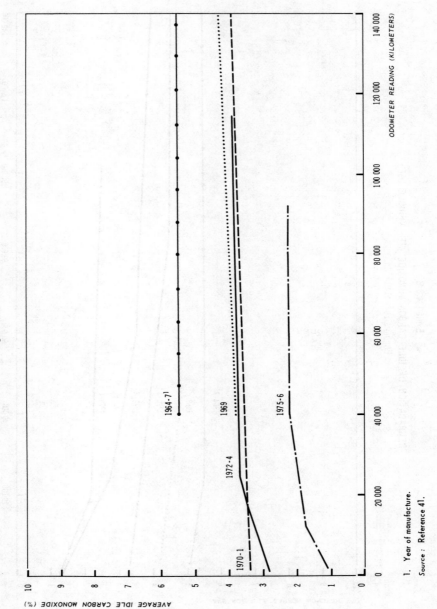

1. Year of manufacture.
Source : Reference 41.

Exhibit 4.8 B

DEGRADATION OF IDLE CARBON MONOXIDE WITH VEHICLE USE

Engine size > 140 c.i.d.

1. Year of manufacture.

Source: Reference 41.

Exhibit 4.9 A

DEGRADATION OF IDLE HYDROCARBONS WITH VEHICLE USE

Engine size 50-140 c.i.d.

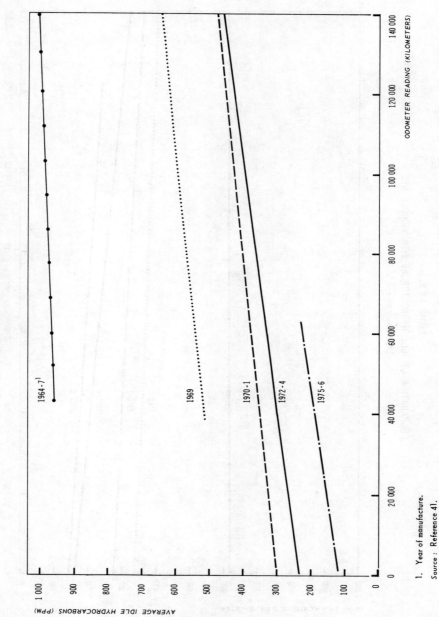

1. Year of manufacture.

Source: Reference 41.

Exhibit 4.9 B

DEGRADATION OF IDLE HYDROCARBONS WITH VEHICLE USE

Engine size > 140 c.i.d.

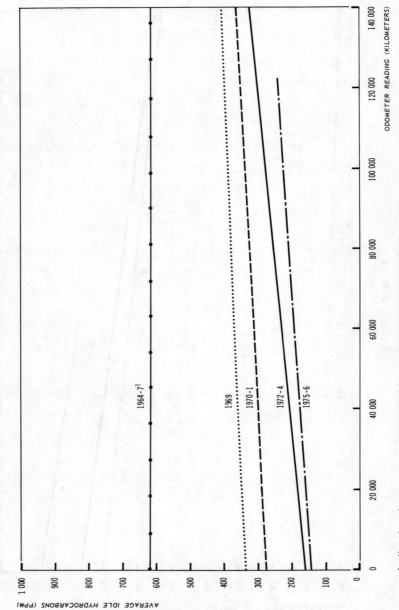

1. Year of manufacture.

Source: Reference 41.

Exhibit 4.10
RELATIONSHIP OF IDLE SCREW POSITION TO IDLE
CARBON MONOXIDE AND IDLE RPM

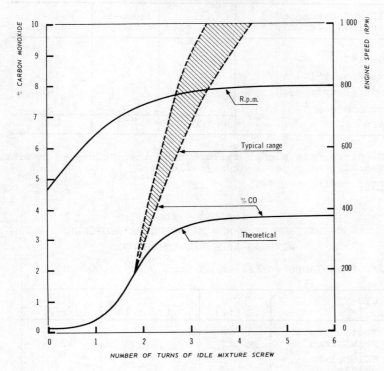

the consumer by making the car idle smoother, have greater pick-up, etc.

It is in these three factors that the solution to improve field performance of emissions will be found.

4.5. OPTIONS FOR IMPROVING EMISSION CONTROL PERFORMANCE IN-USE

As identified in the previous Section, there appears to be three major areas which can contribute to the degradation of controls. Re-stating them they are:

- Control adjustability (and durability);
- Servicing industry ability and propensity to set vehicles to manufacturers' specifications;
- Enforcement of control performance in the field.

Exhibit 4.11

COMPARISON OF OWNER MAINTENANCE RESPONSES
WITH AS-RECEIVED EMISSION TESTS

(Percentage)

"Followed Recommended Service Intervals?"	All Vehicles	Passed CVS-75	Failed CVS-75
"Closely"	74	47	27
"Not closely"(1)	16	6	10
"None" (required)	7	3	4
No Response	3	0	3
Totals	100	56	44

1) Significantly higher failure rate than "closely" maintained vehicles at 5 per cent significance.

Source: Reference 44.

Exhibit 4.12

COMPARISON OF OWNER MAINTENANCE RESPONSES
WITH AS-RECEIVED EMISSIONS

(Non-Tampered Passenger Cars Over 10,000 Miles)

CVS-75	HC (g/km.)	CO (g/km.)	NOX (g/km.)
"Not closely" maintained	0.65	9.09	1.17
"Closely" maintained	0.53	5.72	1.14
Net difference	(1)	-37%	(1)

1) Not significant at 10 per cent.

Source: Reference 44.

4.5.1. <u>Improve Durability or Decrease Adjustability of Control Systems</u>

Several juridictions are considering the use of improved technology to decrease the variation in control levels. In many cases, it is a relatively simple thing to improve the durability of individual components, for instance, improving the life expectancy of diaphragm valves for spark advance on ignition systems. However, it is one thing to say that it can be done and quite another to say how you go about achieving that design goal through means of test standards or production restrictions. There appear to be three major methods to achieve improved durability or decreased adjustability. These are briefly discussed under the following headings.

Change Certification Procedure

If durability of the controls is a problem, then the durability can be improved by increasing the stringency of the durability requirements during certification. In the case of the ECE tests, this would mean imposing durability requirements as they are not required at the present time. In the United States and Japanese programmes, it could be achieved by decreasing the allowable maintenance during the certification programme or extending the mileage over which the durability had to be proved.

Both these ideas have been proposed in California and are being investigated on a nation-wide basis by EPA in the United States. The cost of the decreased maintenance requirements in California has been estimated to produce a net cost saving to the consumer. An example of the effect of durability requirement in California is the standard use of electronic ignition which will decrease the number of ignition tune-ups required during 80,000 km.(*) The ability of vehicles to sustain low maintenance over 80,000 km is directly attributable to the use of a superior grade of gasoline which is low in lead content or, in the case of the United States, lead-free.

The certification programme in the United States as it stands now has achieved several improvements in the maintenance requirements during the 80,000 km certification durability. This is shown in Exhibit 4.13 which is based on General Motors service requirements. While the decrease in these maintenance requirements may improve emissions (although it is not clear that the maintenance requirement is the cause of emission maladjustment), the industry is very strongly opposed in general to the complete elimination of the maintenance requirements. They feel that the maintenance requirements are in place not only for emissions but also, as GM points out,(45) "other more important reasons for recommended maintenance are vehicle safety, fuel economy, reliability, driveability, durability and overall customer satisfaction with the operation of the vehicle".

Independent of whether California or EPA are successful in achieving additional decreases in the amount of maintenance allowed during certification, the fact still remains that there are negligible amounts of maintenance allowed now and there are still component failures and maladjustments in the field. It is thus highly questionable whether this increase in certification severity will have a major effect as many of the components which fail appear to

---

*) In 1975, Chrysler Corporation certified several of its engines for 50,000 miles without a spark plug change using electronic ignition.

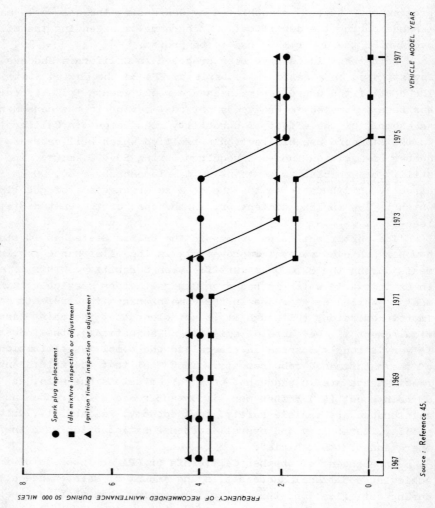

Exhibit 4.13

EMISSION CONTROL MAINTENANCE FREQUENCY VS VEHICLE MODEL YEAR

Source: Reference 45.

be more related to aging or cycling activity than to mileage accumulated at a high rate as in the case of the durability test. For this reason, the approach of attempting to decrease the adjustability of components may prove to be more effective.

Decrease Adjustability

Many of the emission control devices are not adjustable, e.g. EGR or vacuum advance. However, there are two components which have a significant effect, if not the major effect, on emissions which are adjustable: the carburetor and the basic ignition timing. There are two schools of thought on the best way to legislate the reduced adjustability of these components. The first is to limit the maximum adjustment which can be achieved. This is the approach taken in the ECE countries with the 4.5 per cent maximum idle CO. A similar approach could be taken with the basic timing setting, however, this has not been done as of yet. The second approach would demand that the vehicle comply with emission standards regardless of the component setting on the vehicle. This is much the same as the maximum idle CO standard except that it is more stringent because the vehicle must perform a duty cycle under the varied settings. Both of these approaches would decrease the field emissions and be relatively cheap as long as they required only simple design modifications to be made.

Individual Component Specifications

The final method of assuring improved durability in the field is for the control agency to specify the design requirements of individual components in terms of cycle life, temperature or altitude susceptibility, efficiency of operation, etc. Examples of this strategy would be:
- the requirement for enclosed EGR piping (to avoid the deliberate plugging of the hoses);
- antifouling EGR design requirements for leaded-fuelled vehicles; or
- the institution of maximum carburetor per cent CO limits which were "tamperproof".

The most apparent problem with this approach would be the tremendous bureaucracy that it would create. In addition, the pressure on the industry to become more uniform would decrease the diversity of design and perhaps decrease the advancement of technology. On the positive side, it could eliminate certain problems that do arise in the present cars which are related to inadequate material standards.

Before leaving this Section it should be noted that EPA has suggested introduction of the concept of driveability specifications. This has developed from the hypothesis that adjustments of vehicles outside of manufacturer's specifications in the field are to improve the vehicle's driveability. This driveability assessment programme could be as expensive as the original certification programme and may not, again because of the rapid mileage accumulation of such a programme, be exactly representative of what happens to vehicles when in consumer use. As the industry is generally very sensitive to driveability problems, if regulations are in place that made it difficult for them to de-tune a vehicle to supply improved driveability then they would probably be quick to react in improving designs which would meet both the emission levels and supply the driveability that is required by the customers.

4.5.2. Improved Vehicle Servicing

It has been pointed out(46) that there are very few mechanics adequately trained for emission control repair. This fact may be responsible for the poor performance of emission controls although it is more likely that the de-tuning of the vehicles is an overt act. Certainly, in many cases, there is enough cognition of the control systems to de-activate them as illustrated in Exhibit 4.4. Assessments of the mechanic's ability to repair cars that have been identified as being high polluters have been carried out in California to assess the effectiveness of an inspection and maintenance programme(47). They found the service industry lacking in technical ability to diagnose and repair vehicles. It should be remembered that, especially in the United States, mechanics are not always licensed and even where a licensing programme is in place, it may not specify knowledge on emission controls and certainly would not normally require updating or re-testing of mechanics on a regular basis. Thus, it is not at all certain that the service industry can isolate and repair control problems.

There are several ways of improving this ability in the service industry. These can be briefly outlined as follows:

- Improve the serviceability of vehicles - by decreasing the requirement for special tools or equipment and making the setting of the control systems and diagnosis of the state of the control system more easily apparent to the mechanic and thus making the proper repair of the car more likely.
- Improved information to the service industry - although there are several companies in business of providing service specifications to the industry, in many cases these publications have erroneous data in them or data which is not specific enough to any one vehicle. In particular, these manuals are generally weak on idle CO specifications because this is not generally specified by the manufacturer on the car.

- Improved diagnostic technology - by supplying the industry with devices which will provide them with "go - no go" evaluations of control systems and settings, the likelihood of erroneous diagnosis of control failure would decrease.
- Up-graded training of mechanics - although the knowledge of mechanics may be weak, it is not clear that this is a reason for maladjustment. In many cases, it is apparently consumer demands for improved driveability or the most cost-effective, from the mechanics point of view, repair procedure which causes the maladjustment.

Although all of these points could be used to improve the ability of the service industry to repair vehicles, there is still missing the vital requirement of incentive to keep emissions in proper state of repair. As there is currently very little requirement for either the consumer or the mechanic to ensure that the emissions from the vehicle are kept within standards, the natural market forces create a car fleet which is basically de-tuned. It is doubtful that the majority of people will take individual action against air pollution as the benefits of air pollution are spread across all of society and thus there is low perceived individual benefit from vehicle control. Thus, it would seem that either more emphasis on the enforcement of control levels or the removal of the "temptation" of control maladjustment or de-activation would be more suitable courses of action to take to ensure proper control levels on in-use vehicles.

4.5.3. <u>Increased Field Enforcement of Emission Control Standards</u>

The last strategy that could be used to attempt to improve field performance of emission control systems is to increase the enforcement of the standards. Basically, this strategy increases the risk of a penalty to the owner of having a malfunctioning control system. There are a number of ways that this increased enforcement can be carried out as shown in Exhibit 4.14. There are 3 main methods of measurement:

- Engine parameter measurement - this would be the actual testing of individual engine components such as EGR operation or vacuum advanced operation;
- Exhaust concentration measurement - by the simple measurement of exhaust gas concentration a significant amount of information can be gathered concerning the state of the control systems;
- Exhaust mass measurement - this would be similar to the certification test where the actual mass emissions of the vehicle are tested and compared against the original standard.

Exhibit 4.14
INSPECTION TEST OPTIONS

Method of Measurement	Operating Cycle
Engineer Parameter Measurement	Idle/Fast Idle
Exhaust Concentration Measurement	Combination of Steady State Speeds (15/45/100 km/hr)
Exhaust Mass Measurement	Dynamic Driving Cycle (LA-4 Cycle)

Exhibit 4.15
COMPARATIVE ANNUAL OPERATING COSTS

Annual Costs (1976 US$)	Single Lane		Double Lane	
	Idle	Loaded	Idle	Loaded
Capital	13,585	14,215	16,580	17,830
Labour	71,750	71,750	117,450	117,450
Miscellaneous	7,600	7,600	8,600	8,600
Total	92,935	93,565	142,630	143,880

Source: Reference 49; Cost per year for one testing lane in an inspection station.

In addition to the methods of measurement, there are three distinct forms of tests which could be used for inspection purposes:

- Static test - this is either with the engine at a constant speed such as idle or fast idle or in the case of some parameter measurements, with the engine off idling;
- Modal test - similar to the ECE or Japanese cycles but only testing at the constant speed plateaus.
- Dynamic driving cycle - similar to the United States driving cycle.

There could be two objectives to a field enforcement programme. One is to enforce the standard by testing the vehicles to determine if they meet or do not meet the standard under which they were produced. Secondly, a field enforcement programme could have as its aim only an improvement in the emission's performance of the vehicles. The former has proved in the United States to be a difficult task due to problems of vehicle hardware condition and subtle legal nuances of the original Act. The latter approach can be applied in a much less legally and technically demanding way and can be applied differentially to problem areas or types of cars. Based on this potential benefit only the latter approach is considered for further analysis.

A periodic inspection of motor vehicles for exhaust control has been the subject of a large number of technical reports(48). In attempting to distill the information contained in these reports it has been necessary to avoid many of the minor technical arguments that would be contained in a more detailed analysis of inspection and maintenance.

The development of inspection procedures has identified two major tests that are generally considered to be feasible in the field. They are the idle test and the loaded test. The idle test is more simplistic and requires less instrumentation and, in particular, does not require a dynamometer. However, the structural costs associated with both programmes and the data acquisition costs are similar. In a study of the cost of these two alternatives in Nevada(49), the cost shown in Exhibit 4.15 were developed. It can be seen that there is relatively little difference between the operating costs of these two programmes.

The effectiveness of inspection maintenance also varies with the type of test that is used. In studies carried out by the California Air Resources Board(50), the reduction in emissions shown in Exhibit 4.16 were developed. These data indicate that the idle test was marginally better at reducing emissions (as measured on a full CVS-CH test) then were the loaded tests. This difference was a function of both the service industry's ability to repair and the test's ability to identify the high emitters. In data collected in a variety of programmes (shown in Exhibit 4.17), the average repair cost for the two test options were statistically different. The California Air Resources' Riverside Project (51) indicated that the loaded test group cost on average $2 more per repair. The California engineers concluded:

> "What this all seems to indicate is that the repair industry was inadequately trained to handle the extra information from the loaded mode inspection and maintenance programme. The loaded inspection and the idle inspection appear to be equally good in detecting gross emitters which are in need of emission related repairs, but the main advantage of the loaded inspection is in the additional diagnostic information it provides. Based on this study, the repair industry in Southern California does not appear to be able to use this extra information at this time. Not only does the information seem to be unused, but it also seems to generate confusion which can result in slightly higher repair costs. Furthermore, the special training and equipment inspections provided at the beginning of the surveillance programme would be needed throughout the repair industry simply to achieve the effectiveness shown in this study."

Exhibit 4.16
INITIAL EFFECTIVENESS OF EMISSIONS MAINTENANCE

	Hydrocarbons		Carbon Monoxide		Nitrogen Oxides	
	Loaded	Idle	Loaded	Idle	Loaded	Idle
Pass (g/km)	2.36	2.50	26.33	25.91	1.85	1.97
Fail (Before repair, g/km)	4.13	4.48	43.31	40.66	1.74	1.81
Fail (After repair, g/km)	2.67	2.77	27.33	27.31	1.81	1.73
Per cent decrease (Failed Vehicles)	35.54	38.19	34.06	32.84	-4.33	4.20
Per cent decrease (Total Fleet)	17.25	18.76	15.62	15.04	-1.46	1.39

Source: Reference 50.

Exhibit 4.17
AVERAGE COST OF REPAIR PER VEHICLE

Programme	Test Method	
	Loaded	Idle
Arizona	$25.42	
California	$22.81	$20.65
New Jersey		$32.97
Oregon		$18.86

Source: Reference 48.

Thus it would seem that for in-use inspection, the most cost-effective test would be the idle test. However, it must be borne in mind that the CO idle test gives a very limited information on the state of the control systems and only for one of the pollutants. One cannot exclude the possibility that the exclusive use of the idle test could influence on the design of engines with focus on that specific test, disregarding the emission of other pollutants, especially NOx. Thus, there will still be the need for a test procedure less complicated and less costly than the certification test - hopefully without the use of a dynamometer - and yielding results that give full information on the state of the control systems and technical ability to diagnose and repair vehicles.

By using the idle test on the entire population of vehicles with a reasonable test standard, it is expected that approximately a 15 per cent decrease in carbon monoxide and a 19 per cent decrease in hydrocarbons (refer to Exhibit 4.16) could be achieved at the time immediately following repair. The durability of this decrease has been shown in numerous studies to be approximately one year, i.e., after one year any benefit has been eliminated due to component degradation. Thus, the annual overall effectiveness is approximately half the immediate effectiveness of the inspection maintenance programme. Coupling this relatively low effectiveness to a cost of inspection of approximately $2 and an average repair cost in the order of $5 to $8, it can be seen that over the lifetime of the vehicle (ten years) the cost of such a programme would approach $100 per vehicle. This cost must be traded off against the alternatives available to achieve improved in-use emission control.

The selection of inspection standards plays a critical role in the effectiveness of an inspection/maintenance programme. If a uniform standard is applied it will be set at a level such that the highest per cent CO design will pass. This will cause a variable control requirement to be imposed on vehicles; e.g., if the standard was 4.5 per cent CO then a vehicle whose specification was 1.5 per cent CO would have higher emissions than one whose specification was 3.5 per cent CO. An inspection/maintenance system which used a variable standard which was associated with the manufacturer's specification would have a higher effectiveness without a large increase in programme cost.

In general, the software associated with inspection/maintenance is still developing along with the advances in the hardware for testing of vehicles and repairing them. As this progression develops, it is to be expected that the overall cost effectiveness of inspection/maintenance will improve.

A major negative aspect associated with inspection/maintenance is its requirement to incur direct government expenditures on a scale much above those required for regulatory activities. Although by the use of contractors, this funding can in part be displaced to the private sector, it is still a relatively massive amount of money that is required to be spent. Thus, the programme generally has difficulty becoming established because of restraint pressures within government. For this reason, it is generally more attractive for governments to opt for increasing the regulatory aspect of production as opposed to the regulatory aspect of vehicle use.

5. CONCLUSIONS

On the basis of the foregoing review, it is apparent that a major decrease in the effectiveness in emission control systems occurs because of degradation of the system in consumer use. As a result, the ECE legislation is not sufficient by the time the vehicle has accumulated 80,000 Km and the United States 1975 standards are equivalent to the original 1973 standards. The cost analysis indicates that if component adjustability limits were imposed there is the possibility of no net cost because the original cost is expected to be offset by service cost savings by the owner. Thus, this strategy receives the highest cost-effectiveness rating of all possible courses of action. It is thus recommended that major efforts be initiated to introduce restrictions on the adjustability, durability or maintenance requirements of critical emission components.

The cost-effectiveness of the adoption of more stringent emission standards is variable among the countries reviewed. ECE is still at a low level of control and therefore can increase their standards stringency with the lowest marginal cost. In fact, the lowering of the standards may stimulate the application of some technology not commonly used in Europe which could lower the maintenance costs, e.g., lead-free fuel, high energy and electronic ignition. Thus, there is a distinct possibility that these maintenance savings will nullify any cost increases. Special attention might be paid in the ECE to combining the adjustability reduction to control CO and HC with continued lowering of the NOx standard. This approach should preserve the current fuel efficiencies yet allow for decreases in NOx which is becoming a vexing problem.

Both the Japanese and the Americans have committed themselves to low levels of emissions by the 1980s. In taking this step, they will be adding $35-110 onto the cost of the vehicles or approximately doubling the total control cost of the 1977 systems. The effectiveness of these controls will be in serious jeopardy if adjustability controls are not on the control hardware as these advanced systems are even more sensitive to engine parameters, specifically air/fuel ratio than the current generation of controls.

In certain critical air quality areas the imposition of inspection and maintenance can be considered even though it has a relatively low cost-effectiveness ratio as currently applied. Its application only in problem areas supports the concept of "polluter pays" by allowing rural areas lower control cost. It thus effectively creates a two-car strategy. Inspection and maintenance should also cause the service industry to improve the quality of its work even on uninspected vehicles. This is especially important consideration if not all the vehicles are required to be inspected in an effort to increase the cost-effectiveness of the measure.

The countries which are not part of the three major standards all are facing the difficult question of what their next standard step should be. The arguments presented for adjustability standards apply to them as well as the major nations. However, if these countries are considering new car standard reductions considerable economic and technical background work will be required in order that a factually based decision can be made. This regulation development may severely tax the technical resources of the individual governments. It is suggested that because all the countries have a common problem that they pool their knowledge and resources and co-operatively search for the solution. As a minimum effort, a joint technical meeting for information transfer could be extremely useful.

ECE is in a similar position of having to decide on new standards soon. Additionally, they must decide if their test method is valid for lower standards and, in the light of recent data by the CCMC on driving characteristics whether the driving cycle should be changed. This report also highlighted the lack of durability requirement on the certification cars, a provision that has proved to be of some importance in the United States and Japan.

The ECE has historically relied upon the auto industry and the individual nations for the technical and economic appraisal of standards proposals. As it has been seen in this analysis that the industry cost estimates are higher than government estimates, the biasing of analysis by industry must not be ruled out. In addition the uncoordinated analysis by governments makes the level of engineering and economic assessment that is typical in the United States unattainable. It is therefore recommended that co-operative research and assessment programmes be sponsored within the ECE (or EEC) to increase the depth of policy analysis by the member states.

Unlike both the United States and Japan, the European standards are completely uniform and thus there is no jurisdiction (United States) or vehicle designation (Japan) which is used as a "test bed" for new standards. In both Japan and the United States, this policy has appeared to have significant benefits both from the control reliability and cost points of view. It is possible that similar benefits would be present if the European community used one small market portion as their "test" of the control system before they are applied uniformly across all of Europe. Because of the difference in vehicle size, fuel types and test methods, this trial testing will not just duplicate experiences in the United States and Japan but rather will develop uniquely European data which can be used by the manufacturer's to optimise their final large production run vehicles.

Europe should also be considering the application of evaporative emission control standards as this standard has been proven to be very cost-effective in both the United States and Japan. The absence of evaporative controls application in Europe is another example of the lack of technology transfer from the United States and Japan to Europe.

In both the United States and Japan, major efforts to further quantify the durability of the catalyst control systems should be carried out. In particular, the development of a fast check on catalyst activity is an obvious research goal. It should be stressed that the aim of these field programmes should be not to assess whether the cars meet the standards, but to develop and implement systems to improve the effectiveness of the controls in use (e.g. "go-no-go" catalyst test). In conjunction with this research, appraisal and costing studies of low adjustability engine control systems are required in order to sustain the effectiveness of the originally installed control equipment.

Some consideration could be given to the formation of an international advisory council on emission control systems. This group could suggest strategies which are suitable and applicable to developing nations whose technical assessment capabilities are still in their infancy while their vehicle emission problems are becoming as serious as those of the developed nations. Because of the extended age of the vehicles used in these developing countries, and the lack of skilled mechanics and parts, the optimal control strategies are likely to be considerably different than those applied in the advanced nations.

Finally, the selection control strategies, hardware and abatement timetable is very much a political decision which must by definition vary from global region to region. Although the politics of the controls may vary, every attempt must be made to supply the technical analysts with data representative of the state-of-the-art so that every possibility of an uninformed and imprudent decision is excluded.

LIST OF REFERENCES

1. US Congress, Clean Air Act Amendments of 1970 PL 91-604, Washington, D.C.
2. US EPA, Comparison of Japanese and US Automotive Emission Standards, Office of Air and Waste Management, Washington, 1977.
3. OECD, Environmental Implications of Options in Urban Mobility, Environment Directorate, Paris, 1973.
4. Umweltbundesamt, Recommendations to Reduce the Emissions of Pollutants - Motor Vehicle Exhaust Emissions, Federal Environment Agency, Berlin, 1976.
5. Volkswagenwerk AG, Measuring and Testing Procedures for Motor Vehicle Exhaust Emissions, Wolfsburg, FRG., 1977.
6. Data developed from discussions with EPA and Environment Canada staff.
7. Based on data obtained from French, Swiss and German regulatory bodies.
8. US EPA, The Economics of Clean Air, report to US Congress, Document No. 92-67, Washington, 1972.
9. National Academy of Sciences, Report by the Committee for Motor Vehicle Emissions, Washington, 1973.
10. Battelle Memorial Institute, Cost of Clean Air 1974, NTIS PB-238 762, Springfield, Va, 1974.
11. National Academy of Sciences, Report by the Committee on Motor Vehicle Emissions, Washington, 1974.
12. US EPA, Automobile Emission Control - The Current Status and Development Trends as of March 1976, Washington, 1976.
13. General Motors Corporation, Published estimated costs, 1977.
14. US EPA, An Analysis of Alternative Motor Vehicle Emission Standards, Washington, 1976.
15. US EPA, Trade-offs Associated with Possible Auto Emission Standards, Washington, 1975.

16. International Permanent Bureau of Motor Manufacturers (BPICA), <u>Increases in Cost and Fuel Consumption of Passenger Cars Resulting in CO and HC Emission Limits as from 1980</u>, Geneva, 1976.

17. International Permanent Bureau of Motor Manufacturers (BPICA), <u>Increases in Cost and Fuel Consumption of Passenger Cars Resulting from Reductions in CO, HC and NOx Emission Limits as from 1980</u>, Geneva, 1977.

18. Committee of Common Market Automobile Constructors (CCMC), <u>Study on the Evaluation of the Cost/Effectiveness Ratio of Antipollution Systems Installed on Cars</u>, 1977.

19. Supra. 4.

20. German representative's comments at OECD meeting in Paris, Nov. 1977.

21. Supra. 12.

22. Supra. 15.

23. Austin, T.C., Michael R.B., and Service, G.R., <u>Passenger Car Fuel Economy Trends Through 1976</u>, SAE No. 750957, Warrendale, Pa., 1975.

24. Supra. 15.

25. Supra. 14.

26. Supra. 16.

27. Gray, D.S., Azhari, A.G., <u>Saving Maintenance Dollars with Lead-free Gasoline</u>, SAE No. 720084, Detroit, 1972.

28. Supra. 14.

29. Supra. 9.

30. Supra. 4.

31. Data obtained by OECD from FORD of Europe.

32. EPA, <u>Automobile Exhaust Emission Surveillance Reports</u>, annual for 1972, 1973, 1974, Mobile Sources Division, Washington.

33. EPA, <u>Compilation of Air Pollutant Emission Factors</u>, Office of Air and Waste Management, Research Triangle Park, N.C., 1977.

34. Gunderson, J.A., Resnick, L.J., <u>Degradation Effects on Motor Vehicle Exhaust Emission</u>, SAE No. 760366, Detroit, 1976.

35. Appleby, M.R., Bintz, L.J., Tappenden, T.A., <u>Exhaust Emission Levels of In-Use 1975-1976 California Automobiles</u>, SAE No. 770169, Detroit, 1976.

36. Data supplied to OECD by the Swedish government.

37. Gibbs, R., Wotzak, G., et al., *Emissions from In-Use Catalyst Vehicles*, SAE No. 770064, Detroit, 1977.

38. California Air Resources Board, *1975-1976 Model Year Surveillance Test Program Report*, Vehicle Surveillance Section, El Monte, California, 1977.

39. EPA, *Automobile Emission Control-Technological Approaches Towards Improving In-Use Vehicle Emission Performance*, Emission Control Technology Division, Ann Arbor, 1976.

40. Motor Industry Research Association (MIRA), *In-service Emissions of Cars Manufactured to Meet ECE Regulation 15*, Nuneaton, England, 1976.

41. Data received from the Ontario (Canada) Ministry of the Environment, 1977.

42. New Jersey Department of Environmental Protection, *New Jersey Motor Vehicle Emission Inspection Programme Summary and Report Phase I*, 1976.

43. Voelz, E. L., Coleman, E.C., Segal, J.S., Gower, B.G., *Exhaust Emission Levels of In-Service Vehicles*, SAE No. 720498, Detroit, 1972.

44. Supra. 38.

45. General Motors Corporation, *Statement to California Air Resources Board on Proposed Changes to Allowable Maintenance Practices*, San Diego, California, 1977.

46. Committee on Motor Vehicle Emissions, *Field Performance of Emissions-Controlled Automobiles*, report to EPA, Washington, 1974.

47. Bureau of Automotive Repair, *California Vehicle Inspection Program, Riverside Trial Program Report*, Riverside, California, 1977.

48. N.D. Lea & Assoc., *A Study and Compilation of Policies and Data Relevant to In-Use Inspection of Vehicles for Emission Performance*, Prepared for Mobile Source Division, Environment, Canada, Ottawa, 1977.

49. Castaline, A.H., *Inspection and Maintenance in the Nevada Context*, Motor Vehicle Emission Control Conference, Hyannis, Mass., 1976.

50. California Air Resources Board, *Evaluation of Mandatory Inspection and Maintenance Programs*, Sacramento, California, 1976.

51. Supra. 47.

52. Data extracted from material presented to OECD by Japanese Delegation in report entitled "Recent Counter Measures for Air Pollution Control in Japan" (1977).

53. Japanese Environment Agency, Quality of the Environment in Japan, 1977, Tokyo, Japan, 1977.

54. Japanese Environment Agency, Air Pollution and Motor Vehicle Emission Control in Japan, Air Quality Bureau, Tokyo, Japan, 1977.

55. Japanese Environment Agency, Motor Vehicle Emissions, Final Report on Motor Vehicle Nitrogen Oxides Emission Control Technology, Tokyo, 1976.

Appendix

EMISSION TEST CYCLES AROUND THE WORLD

by Dennis J. Simanaitis, Associate Engineering Editor
reprint from Automotive Engineer's Journal, 1977

Cycles used in emission-control certification have profound influence on the design process of today's - and tomorrow's - automobile. Here is a review of test cycle philosophy, design, and operation, as practiced in the United States, Japan and Europe.

Emission-control test procedures do more than merely certify that a car meets some country's set of standards. In a very real sense, they also react as mirrors of driving conditions around the world. And, in a curiously roundabout way, test cycles have profound influence in the vehicle design process itself.

The 1975 Federal Test Procedure currently in use traces its genesis to a trip around downtown Los Angeles. Its foreign counterparts differ philosophically from the FTP, but each in its own way serves to quantify what are perceived as indigenous driving patterns. The matter of cycle-to-cycle correlation is a complicated one; indeed, it appears formidable enough to correlate results within a given cycle. Nevertheless, these test procedures used in certification around the world have become essential features in the realities of automotive design and development.

FTP PROGENITORS

Studies of auto emissions began in this country during the middle 1950s. The Automobile Manufacturers Association, predecessor of MVMA, surveyed traffic patterns in the Los Angeles area and formulated an 11-mode cycle representative of typical countywide driving conditions. This cycle was used in 1956 to evaluate HC and CO baselines from a sample of some 169 vehicles. A Field Service Panel of the Co-ordinating Research Council ran these tests.

In 1959, California's Department of Public Health entered the picture with procedures for testing emission-control devices; they too used the 11-mode cycle. Over the next five years, California's Motor Vehicle Pollution Control Board modified the cycle into its ultimate 7-mode form shown in Fig. 1. California certification of

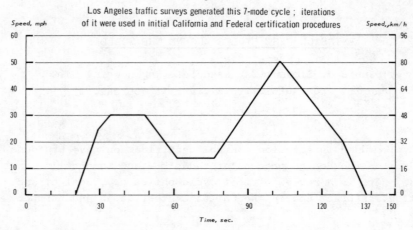

Figure 1

Los Angeles traffic surveys generated this 7-mode cycle; iterations of it were used in initial California and Federal certification procedures

cars built during the 1966 model year made use of this cycle. It can be noted that the 7-mode - not unlike current European and Japanese counterparts - is a somewhat artificial composite of driving modes, rather than a simulation of some actual road route.

This last point is noteworthy, for in 1964 fundamental changes in philosophy began to develop. First, the 24 hour countywide setting gave way to a worst-case modelling of downtown Los Angeles during the morning rush hour. Second, a collage of characteristic driving modes was to be replaced by a simulation of an actual road route. After several equipment-limited false starts - indeed, data acquisition and reduction continue to be thorny problems - researchers settled on what has come to be known as LA 4. A map of this downtown Los Angeles route, together with some representative speeds, is shown in Fig. 2.

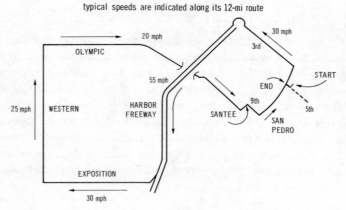

Figure 2

LA 4 was progenitor of current test cycles; some typical speeds are indicated along its 12-mi route

Federal involvement dates from this same period. Iterations of the 7-mode were incorporated into initial Federal procedures used for certifying 1968 automobiles. At the same time, HEW researchers participated in the next generation of LA 4 offspring. The Los Angeles route yielded information which, in turn, formed the basis for similar routes in Cincinnati and Ypsilanti. Out of these studies came EPA's Urban Dynamometer Driving Schedule, and the UDDS became the operating cycle for 1972's Federal Test Procedure.

COMPLEX TRANSLATION

At that time, this translation of road-route driving patterns to dynamometer schedules was a formidable task; indeed, such translation continues to be so. First, the route's driving modes had to be quantified in some reasonable way, without losing subtleties of typical operation. Initially, researchers attempted to identify the accumulated time spent in some 19 driving modes, as measured by engine speed and manifold vacuum. These initial studies were equipment limited, though, for no means existed to keep track of the chronology of events. Later, multi-strip chart recorders gave this capability, but with considerable problems of reducing the mass of data accumulated. (At one point of later EPA cycle development, some 460 ft. of recorder chart lined the halls of EPA's Ann Arbor facility.) Another attempt made use of a fifth wheel combined with strip recording, but it too suffered from problems of data reduction.

Because of equipment limitations, the whole matter was temporarily shelved in 1963. It was not until the late 1960s that data acquisition and reduction techniques were up to the task of measuring speed-time relationships of a selected route and translating them into a dynamometer schedule. Even today, better representations of routes accompany each improvement in data-processing methodology.

Despite these complexities of genesis, a road-route dynamometer cycle has much to offer. It is, after all, a model of some actual vehicle usage. Thus, a well-designed, non-repetitive cycle can contain many nuances of actual driving, characteristics often discarded in iterations of more artificial driving cycles. There are, of course, trade-offs of time, both in designing the cycle and its execution. Also, care must be taken to ensure that the road-route is traffic limited - and not vehicle limited. Otherwise the problems of identifying characteristic modes among several test vehicles become difficult indeed. In any event, there appears to be consensus that road-route-generated cycles offer a most reasonable way of quantifying actual driving conditions.

In developing the UDDS, EPA engineers retained many of LA 4's patterns. By selectively deleting several repeated profiles, they were able to reduce trip length from its original 12 mi to around 7.5 mi, this latter identified in updated Los Angeles surveys as close to the average trip length. In final form the UDDS has an average speed of around 20 mph; its speed-time profile is pictured throughout the article.

The cycle became linked to a procedure which has come to be known as the CVS-C because of its method of collection - constant volume sampling - and its cold start. As such, the CVS-C procedure was used in 1972-74 for purposes of vehicle certification. A major addition came in 1973, with the introduction of a limit for previously unregulated NOx emissions.

1975 FTP

The transition from 1972's procedure to the 1975 FTP in current use was motivated by further research in Los Angeles as well as in other metropolitan areas. The Co-ordinating Research Council's CAPE-10 Project, for example, identified that the UDDS was somewhat idle-heavy and had generally lower speeds than a corresponding five-city composite cycle. Nevertheless, there were no significant differences in emission characteristics. On the other hand, CAPE-10 results suggested the need for a hot-start addition to the procedure. A typical vehicle in the Los Angeles area made approximately 4.7 trips each day: one after an overnight soak - and hence, a cold start - but the others beginning at conditions somewhere between cold and hot starting modes. Also, their evidence suggested that generation of photochemical smog was traceable to broader inputs of time and terrain than had previously been supposed. These observations gave rise to the hot-start portion of the 1975 FTP and, thus, to its full acronym CVS-CH - constant volume sampling - cold, hot.

In developing the dual-mode procedure, EPA engineers had as one option simply running the 7.5 mi 1972 FTP cycle twice, with a hot start immediately following the cold-start phase. This, however, appeared to have several shortcomings. After sufficient warm-up, for example, stabilized cold-start emissions are not significantly different from those produced in a hot-start mode. Thus, two complete runs through the cycle would seem to be time-consuming overkill. Also, two separate calculations would weigh the transient warm-up phases of cold- and hot-start equally; this, despite engineering judgements of EPA that the latter are somewhat more likely. These issues were resolved by devising a procedure based on the 1972 FTP cycle, but with only partial repitition in the hot-start phase and with weighted calculations of transient portions.

Table 1 compares some characteristics of the 1972 FTP with those of its 1975 counterpart.

Table 1
1972 vs. 1975 FTP

	1972	1975
Total Length mi	7.5	11.09
Driving Time. min	22.87	31.3
Avg. Speed. mph	19.7	21.2
Max. Speed. mph	56.7	56.7
Max. Accel. ft/sec^2	4.84	4.84
Time in Mode %		
Cruise	7.9	7.7
Accel.	39.6	39.3
Decel.	34.6	34.9
Idle	17.8	18.1
Stops per Mile	2.3	2.0

A COMPLETE CVS-CH CYCLE

In essential details, a 1975 FTP cycle begins with preconditioning of a chassis dynamometer. It is adjusted for inertia weight and a 50 mph road-load horsepower, based on loaded weight of the vehicle being tested. Ambient conditions of temperature, barometric pressure, and humidity are recorded; temperature is allowed the rather broad range of 20 to 30 deg C (68-86 deg F). The sampling system consists of a positive displacement pump which draws a constant volume of exhaust gases and dilution air into collection bags through a mixing chamber and heat exchanger.

The vehicle is run cycle with emissions collected in three phases. These comprise:

Cold Transient Phase

After starting from cold, the vehicle begins this 505-sec warm-up phase with a 20-sec idle. Speeds as high as 56.7 mph are reached, reminiscent of the Harbor Freeway portion in the cycle's LA 4 antecedent (recall Fig. 2). Four other idle periods are interspersed through this first phase of about 3.6 mi. Its "first-bag" time of 505-sec reflects a compromise between the need for adequate warm-up and a desire to keep total test time within reason.

Cold Stabilized Phase

The second collection period runs through the remaining 868 sec (about 3.9 mi) of the basic driving schedule. This phase has rather more stop-and-go driving than the first, with speeds rarely above 30 mph and some 13 idle periods. Again, those readers who are familiar with the LA 4 route of Fig. 2 may wish to imagine a vehicle's progress; recall, though, that several repeated profiles were selectively deleted. At any rate, the actual trip ended on San Pedro, and the second phase ends after a total of 1372 secs and a simulated trip of about 7.5 mi. Exhaust collection continues for an additional 5 secs after engine shut-down, in the event that any dieseling "run-on" occurs.

Hot Transient Phase

After a 10 min soak at ambient conditions, the vehicle is re-started for the third and final collection period. Idle driving schedule reproduces the first 505 sec. already described.

Emission calculations, in g/mi, are based on the weighted formula $(0.43\ C_t + C_s + 0.57\ H_t)/7.5$, where C_t, C_s and H_t are mass measurements of the given pollutant collected during cold transient, cold stabilized, and hot transient phases, respectively. The weighting factors 0.43 and 0.57 are based on judgements reflecting the nature of typical conditions accompanying the average 4.7 starts each day. It is assumed that two of these could be characterised as cold starts, and the remaining 2.7 could be termed hot. Thus, for example, 2.7/4.7 gives the 0.57 weighting for hot-start data. Also, since stabilized portions of hot and cold starts are essentially equal, only the latter is actually run; this, together with the transient weighting factors, accounts for the 7.5-mi denominator.

This CVS-CH procedure is used in both the United States and Canada for current certification purposes. For the sake of completeness, it can be recalled that 1977 United States limits for HC, CO, and NOx are 1.5, 15, and 2.0 g/mi, respectively. At the time of this writing, 1978 limits still hinge on joint Congressional action.

Canadian emission limits for 1975-1980, obtained with identical CVS-CH procedures, are 2.0, 25, and 3.1 g/mi for HC, CO, and NOx, respectively. They also have a CO-idle test under consideration. Although Canada's emission limits are somewhat less stringent, some 50-60 per cent of Canadian vehicles are built to United States Federal form as a means of achieving commonality with United States counterparts.

In summary, the CVS-CH cycle is a carefully designed simulation of vehicle operating conditions; rather more sophisticated, as will be seen presently, than procedures used in Japan or Europe. It is evident that development and execution of such a procedure are extensive - and expensive - undertakings.

JAPANESE CYCLES

Emission-control certification in Japan, part of the all-inclusive "type test", includes two distinct cycles: a 10-mode procedure - essentially a hot-start urban cycle - and an 11-mode cycle with cold start and somewhat higher speeds. There are also separate evaluations of idle-mode HC and CO, and a 6-mode diesel cycle based on criteria of engine speed and load. As might be expected, contrasting conditions on the road, in politics, and at the industry/government interface have generated procedures which differ considerably from those in this country.

The hot-start urban simulation embodied in Japan's 10-mode cycle is preceded by a warm-up of the test vehicle at a nominal 40 km/h for a minimum of 15 min. After a single-bag collection obtained during the cycle shown in Fig. 3, emissions are analysed on a g/km basis. In general, 10-mode emission limits have characteristically been United States values translated into g/km terms. Distinctions are made, though, between average allowable maxima for a vehicle batch being tested and maximum limits for any given vehicle in the batch. Also, NOx limits depend on vehicle weight, with 1,000 kg chosen as the break point. Table 2 summarises current Japanese limits for the 10-mode procedure. Japan's final limits, which become effective in 1978, differ by lowering the NOx limit to 0.25 g/km batch average and 0.39 g/km vehicle maximum in the 1,000 kg class which makes up much of Japan's national fleet. These more stringent standards are to be deferred until 1981 for cars imported into Japan.

Table 2
JAPANESE 10-MODE LIMITS

HC	CO	NOx	
0.25(0.39)	2.1(2.7)	0.6(0.84)	< 1,000 kg
		0.85(1.2)	> 1,000 kg

All figures in g/km. First in each pair is batch-average maximum, second is vehicle maximum.

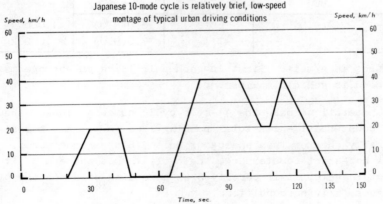

Figure 3
Japanese 10-mode cycle is relatively brief, low-speed montage of typical urban driving conditions

As can be seen from Fig. 3, the 10-mode cycle is a relatively brief, low-speed procedure - just a bit more than two minutes in duration and with speeds no greater than 25 mph throughout. Also, in contrast to the FTP, the cycle makes no attempt at simulating any particular road route; rather, it evaluates emissions as an engine undergoes a brief set of typical urban transient conditions.

Japan's cold-start procedure, the 11-mode cycle, is preceded by a six-hour soak at an ambient temperature between 20 and 30 deg. C. Collection begins with load-free operation for 25 sec, followed by four iterations of the driving schedule shown in Fig. 4. Emission limits for the 11-mode test are characterised in terms of g/test values, with similar batch/individual and vehicle weight distinctions as those in the 10-mode procedure. Table 3 summarises these 11-mode limits.

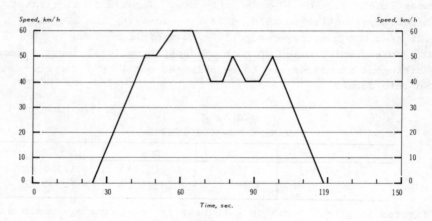

Figure 4
Japanese 11-mode cycle is iterated four times in cold-start procedure

Table 3
JAPANESE 11-MODE LIMITS

HC	CO	NOx	
7.0(9.5)	60(85)	6.0(8.0)	<1,000 kg
		7.0(9.0)	>1,000 kg

All figures in g/test. First in each pair is batch-average maximum, second is vehicle maximum.

Its iterations make the 11-mode cycle somewhat longer than its 10-mode counterpart - its total running time is 8 min - but still considerably shorter than analogous United States procedures. Also, like its hot-start counterpart, the cycle is not patterned after any road route; again, it is perceived as something of a montage of typical Japanese driving conditions.

Because of inherent differences between United States and Japanese procedures, it is difficult to identify any precise correlations of test results, despite their often equivalent limits. It has been observed, though, that typical United States vehicles in California trim tend to meet current Japanese regulations. The cold-start 11-mode cycle, for example, appears less demanding perhaps than its United States counterpart: compared to the FTP's 505-sec first bag, the 11-mode's gram/test emission limits tend to lessen the impact of cold-start HC and CO generation.

EUROPEAN CYCLES

Europe's principal cycle is embodied in ECE-15, a regulation promulgated by the Economic Commission for Europe. This organisation is one of four regional economic commissions set up by the United Nations. ECE regulations are formulated by committees representing some 18 member countries; five other non-member European countries, as well as Australia, Canada, Japan, and the United States, are also represented.

Any regulation becomes mandatory in a country, once its appropriate national regulatory body chooses to adopt it. ECE-15 has been adopted by most Commission members, with Sweden a major car-producing exception. Sweden's F-40 standard is similar to the 1973 United States procedure; that is, the 1972 CVS-C cycle with 1973's added NOx requirement. F-40 sets HC, CO and NOx limits of 3.4, 39, and 3.0 g/mi, respectively. (A similar situation prevails in Australia as well.)

ECE-15 incorporates three distinct procedures, including a cold-start driving cycle, an idle CO test, and a crankcase emission evaluation. The driving cycle is preceded by a six-hour soak at ambient conditions, with temperature between 20 and 30 deg. C. Its driving schedule, shown in Fig. 5, is iterated four times with a single-bag collection beginning after a 40-sec warm-up. Emission limits depend on vehicle weight and are characterised by g/test values for a simulated distance of approximately 4 km. Current limits, dating from 1975, are shown in Table 4.

The omission of a NOx requirement is noteworthy, although it can also be noted that an amendment is in the works which would incorporate such a standard: ECE 15/02 would set NOx limits ranging from 10 to 16 g/test. Indeed, although its effective date is still a matter of conjecture, several manufacturers are already building and testing to this proposed requirement. Another amendment, ECE-15/03, would lower emission limits across the board, perhaps by as much as 20 - 30 per cent; this proposal is expected to be a topic of discussion in Geneva later this month.

Table 4
ECE-15 EMISSION LIMITS

Loaded Vehicle Weight, kg	HC	CO
≤ 750	6.8	80
750- 850	7.1	87
850-1020	7.4	94
1020-1250	8.0	107
1250-1470	8.6	122
1470-1700	9.2	135
1700-1930	9.7	149
1930-2150	10.3	162
> 2150	10.9	176

All emission values are g/test, obtained in a driving schedule of approximately 4 km.

As is evident from Fig. 5, ECE-15 is philosophically close to the Japanese cycles already described. It is a composite of vehicle operating conditions, not a route-generated simulation. Also, compared to the FTP, it is a relatively low-speed procedure. Indeed, some more-powerful cars must brake occasionally to remain in the low-speed cruise mode. ECE-15's low speed, of course, tends to minimise NOx generation while exacerbating the problems of HC and CO control.

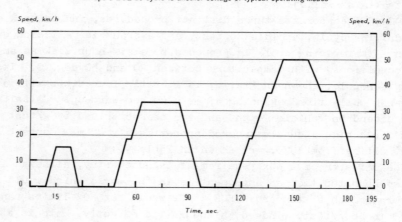

Figure 5
Europe's ECE-15 cycle is another collage of typical operating modes

Once again, no clear-cut correlations exist between ECE-15 results and those, for example, of the CVS-CH procedure. However, it has been observed that typical United States vehicles in Federal trim were comfortably within European limits several years ago.

More recently, United States exports to Europe have traded current Federal levels of emission control with cost, economy, and driveability considerations; they now meet ECE-15 with somewhat closer, albeit still reasonable margins.

A WORLDWIDE CYCLE?

The pros and cons of a uniform worldwide certification procedure form a complex set of tradeoffs. First, consider the advantages. A common procedure, even with differing limits, would be an evident asset from a marketing point of view. Duplication of effort - in some cases, triplication or worse - would be minimised with a single procedure worldwide.

There are design advantages as well. By their very nature, test cycles function both as design paradigms as well as criteria of compliance. Regulators sense that designing to a cycle may well be the most cost-effective strategy for a manufacturer to follow. Thus, given that a cycle is representative of ordinary field conditions, it provides a set of objectives for both manufacturer and regulator. Cold-start controls in this country, for example, are perceived as legitimate engineering responses to the 505-sec first bag of the CVS-CH - and to the evident fact that vehicles do experience cold-starts in actual use. It has been noted, though, that a multitude of regulations tends to promote myriad "fixes" each in response to some peculiarity of a given test-cycle characteristic. Thus, any commonality of procedure could well lead to a desirable design commonality too. The development of Europe's ECE-15, for example, indicates some of the benefits of certification commonality among countries.

On the other hand, compelling arguments, both technical and political, can be offered against a uniform worldwide procedure. In essence, technical arguments revolve around the basic philosophy of simulation: if realistic road routes provide the basis for cycle design, then what chance is there for a "typical worldwide" route? Indeed, it has been suggested that, even in this country alone, a driver in the rural midwest uses his car under markedly different conditions from those experienced by his Los Angeles commuting compatriot.

Also, there would appear to be political subtleties inherent in worldwide uniformity of certification procedures. For a variety of reasons, identical emission limits around the world may not be feasible. Yet consider the case of a country which adopts a uniform test procedure with less stringent limits: would not its environmentalists be in a position to question the wisdom of such a course? It has been observed that a more productive approach may be for each political unit to set its procedures and limits in response to the needs and desires of those governed.

TEST CORRELATIONS

Were accurate correlations realisable from cycle to cycle, some of the shortcomings of differing procedures could be mitigated. However, work done in this direction has yielded only mixed results. Comparing Japan's 10-mode, for example, with the CVS-CH is a complicated undertaking indeed.

As a matter of fact, the problems of maintaining reproducibility within any one cycle appear to be formidable enough. Studies initiated by the National Academy of Sciences have indicated that reliability and reproducibility are strongly influenced by systematic and random variability in test procedures. These influences are crucial, of course, for such information is necessary in setting vehicle design levels which reflect the realities of certification. It has been noted, for example, that some engine families have mean design levels set as low as ¼ of corresponding legal limits.

Both vehicle and procedure have been identified as sources of variability. With the vehicle, it extends all the way from assembly line tolerances to in-service states of tune. On-line emission monitoring and adjustment limiters in carburation and timing have been responses to this problem. Even with back-to-back testing under closely controlled conditions, though, one study showed engine family standard deviations as high as 31 per cent of mean values of measured emission.

Variations in test results have also been studied within a given test cell, from cell to cell, and from lab to lab. MVMA began conducting lab-to-lab cross correlations in 1974 with a specially designed G.M. vehicle, REPCA I. The CVS-CH cycle was run at American Motors, Chrysler, EPA-Ann Arbor, Ford, G.M., and International Harvester facilities. Most of these results fell within two standard deviations of mean values; standard deviations, as percentages of HC, CO, and NOx means, were approximately 4.5, 8.0, and 1.8 per cent, respectively.

Another EPA study indicated that test-to-test variability of composite HC/CO emission was ± 6 per cent, with NOx variability around ± 3 per cent, and CO_2 ± 1 per cent. This good reproducibility of CO_2 data, by the way, was perceived as favourable evidence for using carbon mass balance techniques in calculating fuel consumption figures from cycle-generated data. Indeed, the CVS-CH cycle is used in determining EPA's city fuel economy figures. Highway mpg values are obtained from another cycle, based on EPA's studies of road routes in the Ann Arbor area.

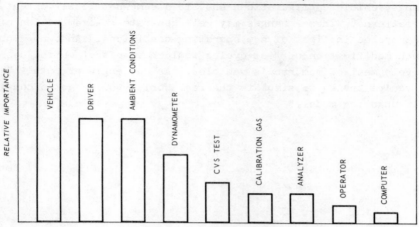

Figure 6
Variables in the CVS-CH procedure : their relative importance

Fig. 6 shows estimates of relative importance attributed to variables inherent in the CVS-CH procedure. The first three, vehicle, driver, and ambient conditions, evidently pertain to any correlation study - a single cell, cell-to-cell, or lab-to-lab. The next five are cell-specific; the last, computer variations, would evidently have to be considered in analysing lab-to-lab correlations.

In summary, the matter of correlation is a complex one. On the one hand, high degrees of reproducibility have evident advantage in vehicle design and development, even if laboratories are somewhat isolated from reality. However, this gives rise to a larger issue of correlating such carefully controlled conditions with those experienced in the field. For example, influences of ambient temperature on emissions are well-known. Yet, should lab conditions be more tightly controlled than the current 68 to 86 deg. F? Or should increased recognition be made of the fact that actual in-service temperatures vary by considerably more than this? The problem is a thorny one, both philosophically and practically, and research continues.

FUTURE TRENDS

Europe's ECE-15/03, if enacted, would narrow the gap between United State's standards and those in Europe. It is also possible that Europe's procedure may adopt some sort of multiple-bag collection, similar to FTP's. Indeed, a recent critique of ECE-15 done by an English organisation, the Transportation and Road Research Laboratory, may provide the basis for discussions bringing European procedures even closer to those used on this side of the Atlantic.

In this country, responses to energy issues have already begun to change vehicle configurations and, to a somewhat lesser extent, usage patterns. These changes may well generate re-examination of current cycles in light of new operating conditions. And, no doubt, any such modification of test cycles would have effect on the design and development of tomorrow's vehicles. As one engineer puts it, "When you're trying to simulate the real world, you've got a constantly changing model."

N.B. The author expresses sincere thanks to officials of the following organisations for providing information in the preparation of this article: British Leyland Motors, Ltd., Canadian Department of Transportation, Ford Motor., General Motors Corp., Toyo Kogyo Co., Ltd., and the Motor Vehicle Emission Laboratory, United States Environmental Protection Agency.

OECD SALES AGENTS
DÉPOSITAIRES DES PUBLICATIONS DE L'OCDE

ARGENTINA – ARGENTINE
Carlos Hirsch S.R.L., Florida 165, 4° Piso (Galería Guemes)
1333 BUENOS-AIRES, Tel. 33-1787-2391 Y 30-7122

AUSTRALIA – AUSTRALIE
Australia & New Zealand Book Company Pty Ltd.,
23 Cross Street, (P.O.B. 459)
BROOKVALE NSW 2100   Tel. 938-2244

AUSTRIA – AUTRICHE
Gerold and Co., Graben 31, WIEN 1. Tel. 52.22.35

BELGIUM – BELGIQUE
LCLS
44 rue Otlet, B1070 BRUXELLES. Tel. 02-521 28 13

BRAZIL – BRÉSIL
Mestre Jou S.A., Rua Guaipà 518,
Caixa Postal 24090, 05089 SAO PAULO 10. Tel. 261-1920
Rua Senador Dantas 19 s/205-6, RIO DE JANEIRO GB.
Tel. 232-07. 32

CANADA
Renouf Publishing Company Limited,
2182 St. Catherine Street West,
MONTREAL, Quebec H3H 1M7 Tel. (514) 937-3519

DENMARK – DANEMARK
Munksgaards Boghandel,
Nørregade 6, 1165 KØBENHAVN K. Tel. (01) 12 85 70

FINLAND – FINLANDE
Akateeminen Kirjakauppa
Keskuskatu 1, 00100 HELSINKI 10. Tel. 625.901

FRANCE
Bureau des Publications de l'OCDE,
2 rue André-Pascal, 75775 PARIS CEDEX 16. Tel. (1) 524.81.67
Principal correspondant :
13602 AIX-EN-PROVENCE : Librairie de l'Université.
Tel. 26.18.08

GERMANY – ALLEMAGNE
Alexander Horn,
D - 6200 WIESBADEN, Spiegelgasse 9
Tel. (6121) 37-42-12

GREECE – GRÈCE
Librairie Kauffmann, 28 rue du Stade,
ATHÈNES 132. Tel. 322.21.60

HONG-KONG
Government Information Services,
Sales and Publications Office, Beaconsfield House, 1st floor,
Queen's Road, Central.   Tel. 5-233191

ICELAND – ISLANDE
Snaebjörn Jónsson and Co., h.f.,
Hafnarstraeti 4 and 9, P.O.B. 1131, REYKJAVIK.
Tel. 13133/14281/11936

INDIA – INDE
Oxford Book and Stationery Co.:
NEW DELHI, Scindia House. Tel. 45896
CALCUTTA, 17 Park Street. Tel.240832

ITALY – ITALIE
Libreria Commissionaria Sansoni:
Via Lamarmora 45, 50121 FIRENZE. Tel. 579751
Via Bartolini 29, 20155 MILANO. Tel. 365083
Sub-depositari:
Editrice e Libreria Herder,
Piazza Montecitorio 120, 00 186 ROMA. Tel. 674628
Libreria Hoepli, Via Hoepli 5, 20121 MILANO. Tel. 865446
Libreria Lattes, Via Garibaldi 3, 10122 TORINO. Tel. 519274
La diffusione delle edizioni OCSE è inoltre assicurata dalle migliori
librerie nelle città più importanti.

JAPAN – JAPON
OECD Publications and Information Center
Akasaka Park Building, 2-3-4 Akasaka, Minato-ku,
TOKYO 107. Tel. 586-2016

KOREA – CORÉE
Pan Korea Book Corporation,
P.O.Box n° 101 Kwangwhamun, SÉOUL. Tel. 72-7369

LEBANON – LIBAN
Documenta Scientifica/Redico,
Edison Building, Bliss Street, P.O.Box 5641, BEIRUT.
Tel. 354429–344425

MALAYSIA – MALAISIE
University of Malaya Co-operative Bookshop Ltd.
P.O. Box 1127, Jalan Pantai Baru
Kuala Lumpur, Malaysia. Tel. 51425, 54058, 54361

THE NETHERLANDS – PAYS-BAS
Staatsuitgeverij
Chr. Plantijnstraat
'S-GRAVENHAGE. Tel. 070-814511
Voor bestellingen: Tel. 070-624551

NEW ZEALAND – NOUVELLE-ZÉLANDE
The Publications Manager,
Government Printing Office,
WELLINGTON: Mulgrave Street (Private Bag),
World Trade Centre, Cubacade, Cuba Street,
Rutherford House, Lambton Quay, Tel. 737-320
AUCKLAND: Rutland Street (P.O.Box 5344), Tel. 32.919
CHRISTCHURCH: 130 Oxford Tce (Private Bag), Tel. 50.331
HAMILTON: Barton Street (P.O.Box 857), Tel. 80.103
DUNEDIN: T & G Building, Princes Street (P.O.Box 1104),
Tel. 78.294

NORWAY – NORVÈGE
J.G. Tanum A/S
P.O. Box 1177 Sentrum
Karl Johansgate 43
OSLO 1   Tel (02) 80 12 60

PAKISTAN
Mirza Book Agency, 65 Shahrah Quaid-E-Azam, LAHORE 3.
Tel. 66839

PORTUGAL
Livraria Portugal, Rua do Carmo 70-74,
1117 LISBOA CODEX.
Tel. 360582/3

SPAIN – ESPAGNE
Mundi-Prensa Libros, S.A.
Castelló 37, Apartado 1223, MADRID-1. Tel. 275.46.55
Libreria Bastinos, Pelayo, 52, BARCELONA 1. Tel. 222.06.00

SWEDEN – SUÈDE
AB CE Fritzes Kungl Hovbokhandel,
Box 16 356, S 103 27 STH, Regeringsgatan 12,
DS STOCKHOLM. Tel. 08/23 89 00

SWITZERLAND – SUISSE
Librairie Payot, 6 rue Grenus, 1211 GENÈVE 11. Tel. 022-31.89.50

TAIWAN – FORMOSE
National Book Company,
84-5 Sing Sung Rd., Sec. 3. TAIPEI 107. Tel. 321.0698

THAILAND – THAILANDE
Suksit Siam Co., Ltd.
1715 Rama IV Rd.
Samyan, Bangkok 5
Tel. 2511630

UNITED KINGDOM – ROYAUME-UNI
H.M. Stationery Office, P.O.B. 569,
LONDON SE1 9 NH. Tel. 01-928-6977, Ext. 410 or
49 High Holborn, LONDON WC1V 6 HB (personal callers)
Branches at: EDINBURGH, BIRMINGHAM, BRISTOL,
MANCHESTER, CARDIFF, BELFAST.

UNITED STATES OF AMERICA
OECD Publications and Information Center, Suite 1207,
1750 Pennsylvania Ave., N.W. WASHINGTON, D.C.20006.
Tel. (202)724-1857

VENEZUELA
Libreria del Este, Avda. F. Miranda 52, Edificio Galipán,
CARACAS 106. Tel. 32 23 01/33 26 04/33 24 73

YUGOSLAVIA – YOUGOSLAVIE
Jugoslovenska Knjiga, Terazije 27, P.O.B. 36, BEOGRAD.
Tel. 621-992

Les commandes provenant de pays où l'OCDE n'a pas encore désigné de dépositaire peuvent être adressées à :
OCDE, Bureau des Publications, 2 rue André-Pascal, 75775 PARIS CEDEX 16.
Orders and inquiries from countries where sales agents have not yet been appointed may be sent to:
OECD, Publications Office, 2 rue André-Pascal, 75775 PARIS CEDEX 16.

OECD PUBLICATIONS, 2 rue André-Pascal, 75775 PARIS CEDEX 16 - No. 40 953 1979
PRINTED IN FRANCE
(97 79 04 1) ISBN 92-64-11913-2